Controversies in Neuro-ophthalmology: A case-based debate

Editors

Andrew G Lee, MD
The Methodist Hospital, Weill Cornell Medical College,
Houston, Texas, USA

Jacinthe Rouleau, MD
Centre hospitalier de l'Université de Montréal (CHUM)
Université de Montréal
Montréal, Québec, Canada

Reid Longmuir, MD
University of Iowa
Iowa City, Iowa, USA

CRC Press
Taylor & Francis Group
Boca Raton London New York

CRC Press is an imprint of the
Taylor & Francis Group, an **informa** business

CRC Press
Taylor & Francis Group
6000 Broken Sound Parkway NW, Suite 300
Boca Raton, FL 33487-2742

First issued in paperback 2019

© 2010 by Taylor & Francis Group, LLC
CRC Press is an imprint of Taylor & Francis Group, an Informa business

No claim to original U.S. Government works

ISBN-13: 978-1-4200-7092-7 (hbk)
ISBN-13: 978-0-367-38472-2 (pbk)

A CIP record for this book is available from the British Library.

Library of Congress Cataloging-in-Publication Data

Typeset by C&M Digitals (P) Ltd, Chennai, India

Visit the Taylor & Francis Web site at
http://www.taylorandfrancis.com

and the CRC Press Web site at
http://www.crcpress.com

Dedication

Dr. Andrew G Lee wishes to dedicate this book to his brother and sister Richard Lee and Amy Lee Wirts MD who taught him the value of debating a controversial topic to the mutual edification of all parties and that people can disagree on subjects without being disagreeable. Dr. Lee thanks his wife Hilary A. Beaver MD (and his two daughters Rachael and Virginia Lee) for tolerating yet another book-writing process. Finally, Dr. Lee wishes to thank his parents Alberto C. Lee MD and Rosalind Go Lee MD for just being great parents.

Dr Rouleau wishes to dedicate this book to her mother for always being the best source of motivation and inspiration, and to all her medical mentors for their teaching and support.

Dr. Longmuir wishes to dedicate this book to his wife Susannah Longmuir MD and his daughter Charlotte.

Contents

Introduction   ix
List of Contributors   xi

1 Should a patient with unexplained isolated optic atrophy have neuroimaging and further laboratory evaluation?   **1**

Pro: A patient with unexplained isolated optic atrophy should have neuroimaging
and further laboratory evaluation   1
Nicholas Volpe

Con: A patient with unexplained isolated optic atrophy does not always need neuroimaging
and further laboratory evaluation   4
Karl Golnik

2 Should a young patient with a new diagnosis of optic neuritis have testing and treatment for multiple sclerosis (MS)?   **7**

Pro: A young patient with a new diagnosis of optic neuritis should have testing and
treatment for multiple sclerosis if identified   7
Fiona Costello

Con: A young patient with a new diagnosis of optic neuritis does not always
require testing and treatment for MS   11
Michael S Lee

3 Should a patient with optic disc edema with a macular star figure (neuroretinitis) have lab testing and treatment?   **13**

Pro: Test for cat scratch, Lyme, syphillis, tuberculosis (Tb) and treat empirically
for cat scratch fever   13
Karl Golnik

Con: Do not test for cat scratch, Lyme, syphillis, tuberculosis (Tb) and do not treat
for cat scratch fever   18
Eric Eggenberger

4 Should a vasculopathic patient with nonarteritic anterior ischemic optic neuropathy have any testing?   **19**

Pro: Test for blood pressure (nocturnal hypotension, 24 hour blood pressure
measurements), sleep apnea, blood sugar, cholesterol, no smoking, aspirin per day   19
Karl Golnik

Con: Do not test for blood pressure (nocturnal hypotension, 24 hour blood pressure
measurements), sleep apnea, blood sugar, cholesterol, no smoking, aspirin per day   22
Michael S Lee

5   **What is the treatment for giant cell arteritis?**   24

Pro: Patients with suspected GCA and vision loss should receive IV steroids followed by
oral prednisone and antiplatelet therapy while awaiting temporal artery biopsy   24
Timothy J McCulley and Thomas Hwang

Con: Oral steroids are adequate treatment for GCA   28
Eric Eggenberger

6   **Should I do a bilateral or unilateral temporal artery biopsy in suspected
    giant cell arteritis?**   29

Pro: A bilateral temporal artery biopsy should be strongly considered in cases
of suspected GCA   29
Michael S Lee

Con: A unilateral temporal artery biopsy is usually adequate   31
Wayne T Cornblath

7   **Should I treat traumatic optic neuropathy?**   35

Pro: Treat traumatic optic neuropathy with high-dose steroid or possibly surgery   35
Nicholas Volpe

Con: There is no proven treatment for traumatic optic neuropathy   38
Eric Eggenberger

8   **Should I do a MRI and MR venogram in every patient with pseudotumor cerebri?**   40

Pro: MRI and MRV are necessary in the workup of possible pseudotumor cerebri   40
Nicholas Volpe

Con: MRI scan with contrast is adequate in the evaluation of possible pseudotumor cerebri and a
venogram is usually unnecessary   43
Fiona Costello

9   **Should we perform carotid Doppler and cardiac echo on young patients
    with transient visual loss?**   45

Pro: Transient vision loss in young people can be thromboembolic so work up
should be performed   45
Nicholas Volpe

Con: Transient vision loss in young people does not routinely require a cardiovascular
risk factor assessment   49
Fiona Costello

10 **What is the best visual field test for neuroophthalmology?**   51

Pro: Goldmann (kinetic) is better   51
Fiona Costello

Con: Automated is better   54
Wayne T Cornblath

11 Does visual rehabilitation therapy help patients with homonymous hemianopsia? 59

Pro: Vision rehabilitation therapy can be useful in cases of homonymous
visual field loss 59
Wayne T Cornblath

Con: Vision rehabilitation therapy is not helpful in cases of homonymous
visual field loss 62
Eric Eggenberger

**12 Should a patient with a pupil involved third nerve palsy have a catheter angiogram
if the MRA or CTA are negative?** 63

Pro: A patient with a pupil-involved third nerve palsy should have an angiogram
if the MRA or CTA are negative 65
Timothy J McCulley and Soraya Rofagha

Con: A negative MRA or CTA is adequate in the evaluation of a pupil-involved third nerve palsy 65
Michael S Lee

13 Do erectile dysfunction agents cause anterior ischemic optic neuropathy? 67

Pro: Erectile dysfunction agents do cause anterior ischemic optic neuropathy 67
Karl Golnik

Con: Erectile dysfunction agents have never been proven to cause ischemic optic neuropathy 71
Timothy J McCulley and Michael K Yoon

14 Does amiodarone produce an optic neuropathy? 74

Pro: Amiodarone does cause an optic neuropathy 74
Eric Eggenberger

Con: Amiodarone does not cause an optic neuropathy 76
Timothy J McCulley and Shelley Day

**15 Should I start my patient with myasthenia gravis on steroids to reduce the chance
of generalized myasthenia gravis?** 79

Pro: Steroids may prevent generalized myasthenia gravis in patients presenting with an
isolated ocular form of the disease 79
Nicholas Volpe

Con: Steroid should not be given to prevent onset of generalized myasthenia gravis 81
Michael S Lee

16 Does radiation therapy work for thyroid ophthalmopathy? 82

Pro: Low dose orbital radiation therapy is a useful alternative in the treatment of
thyroid eye disease 82
Reid Longmuir

Con: Radiation therapy is not helpful in the treatment of thyroid eye disease 84
Karl Golnik

17 **Should I do topical pharmacologic testing in the Horner syndrome?**   87

Pro: Pharmacologic testing is useful in the evaluation of possible Horner's syndrome   87
Fiona Costello

Con: Pharmacologic testing is not necessary in the evaluation of possible Horner's
syndrome and one should proceed directly to neuroimaging   89
Nicholas Volpe

18 **Should a patient with giant cell arteritis have a fluorescein angiogram?**   91

Pro: A fluorescein angiogram is useful in the evaluation of suspected giant cell arteritis   91
Fiona Costello

Con: Fluorescein angiography is usually not necessary in the evaluation
of possible giant cell arteritis   93
Eric Eggenberger

19 **Does pseudotumor cerebri without papilledema exist?**   94

Pro: Idiopathic intracranial hypertension (IIH) without papilledema does exist   94
Timothy J McCulley and Thomas N Hwang

Con: Pseudotumor cerebri without papilledema does not exist   98
Michael Lee

20 **Does a patient with an isolated vasculopathic ocular motor cranial nerve palsy
need a neuorimaging study?**   100

Pro: Despite a clinical diagnosis of ischemic cranial nerve palsy, neuroimaging
should be strongly considered   100
Nicholas Volpe

Con: A clinical diagnosis of ischemic cranial nerve palsy excludes the need
for neuroimaging   101
Wayne T Cornblath

Introduction

Welcome reader to "Controversies in Neuroophthalmology". This text is meant to be a fast, fun read for you. It is intended to present both sides of controversial issues in the management of neuroophthalmic problems. We have assembled a distinguished panel of experts in the field to help us including

Michael S Lee, MD (University of Minnesota)
Fiona Costello, MD (University of Calgary)
Eric Eggenberger, DO (Michigan State University)
Karl Golnik, MD (University of Cincinnati)
Timothy J McCulley, MD (University of California, San Francisco)
Nicholas Volpe, MD (University of Pennsylvania)
Wayne T Cornblath, MD (University of Michigan)

The book is divided into 20 chapters of interest to the general ophthalmologist and each chapter opens with the case that illustrates the controversy and poses a clinical question.

The questions are designed to be open-ended and may have more than one correct answer, thus the controversy. Each expert has been "assigned" either the "pro" or "con" position but the reader should be cautioned that our intent is not to provide hard and fast dogma. Instead, our goal is to provide a balanced viewpoint to a specific controversy from leading experts and then provide a summary statement based upon the editorial consensus of the editors Andrew Lee, MD (Weill Cornell Medical College), Jacinthe Rouleau, MD (University of Montréal), and Reid Longmuir, MD (University of Iowa).

We also wish to state upfront that the opinions expressed in this book are not meant to be construed as a standard of care recommendation and that each opinion is based upon individual but expert practice. This book merely contains reasonable suggestions and alternatives for clinical practice.

We hope that you will enjoy reading this book as much as we enjoyed writing it.

Andrew G Lee, MD
Jacinthe Rouleau, MD
Reid Longmuir, MD

*Disclaimer: The editors and authors of this textbook emphasize to the reader that the contents of this book are meant to promote discussion and debate and are not meant to be interpreted as dogma or an implied or recommended "standard of care". The cases contained are composites meant for teaching purposes only and the opinions of the individual authors and do not represent necessarily the opinions of the editors or publisher. Each clinical controversy is followed by a pro and con position and a final summary statement from the editors. We wish to emphasize that there is not a "right" or "wrong" answer to the questions and the cases are meant to illustrate different clinical approaches to the same clinical problem.

List of Contributors

Wayne T Cornblath
University of Michigan
Ann Arbor, Michigan, USA

Fiona Costello
University of Calgary
Calgary, Canada

Shelley Day
University of California
San Francisco, California, USA

Eric Eggenberger
Michigan State University
East Lansing, Michigan, USA

Karl Golnik
University of Cincinnati and the Cincinnati Eye Institute
Cincinnati, Ohio, USA

Thomas N Hwang
University of California, San Francisco (UCSF)
San Francisco, California, USA

Michael S Lee
The Methodist Hospital
Weill Cornell Medical College
Houston, Texas
University of Minnesota
Minneapolis, Minnesota, USA

Reid Longmuir
University of Iowa
Iowa City, Iowa, USA

Timothy J Mcculley
University of California, San Francisco (UCSF)
San Francisco, California, USA

Soraya Rofagha
University of California, San Francisco (UCSF)
San Francisco, California, USA

Nicholas Volpe
Scheie Eye Institute
University of Pennsylvania School of Medicine
Philadelphia, Pennsylvania, USA

Michael K Yoon
University of California
San Francisco, California, USA

1   Should a patient with unexplained isolated optic atrophy have neuroimaging and further laboratory evaluation?

A 69-year-old man presents to the local ophthalmologist with complaint of poor vision in the left eye. He does not know how long it has been present, but he noticed it upon covering the other eye about one week before. He does not believe the vision has gotten any worse since onset. His past medical history includes hypertension, for which he takes metoprolol, and diet-controlled diabetes. On examination his visual acuity is 20/25 OD and 20/60 OS. There is a 1.5 log unit relative afferent pupillary defect (RAPD) on the left. The slit lamp examination reveals only mild nuclear sclerotic cataracts. His Humphrey visual fields testing shows marked field loss on the left, and a full field on the right (Figures 1.1 and 1.2). He has a pale optic disc nerve on the left compared to the right (Figures 1.3 and 1.4), which is confirmed on optical coherence tomography (OCT) of the retinal nerve fiber layer (RNFL) shown in Figure 1.5.

PRO: A PATIENT WITH UNEXPLAINED ISOLATED OPTIC ATROPHY SHOULD HAVE NEUROIMAGING AND FURTHER LABORATORY EVALUATION

Nicholas Volpe

The vast majority of patients with isolated optic atrophy will turn out to have old, inactive, often never identified, problems such as previous ischemic optic neuropathy, optic neuritis, congenital or hereditary abnormalities, trauma or some other long-standing cause for the optic atrophy. This, however, must be a position of exclusion, particularly if awareness of vision loss is new. The examining physician is obligated in the absence of supporting history or documentation to pursue the possibility of a treatable cause for the optic nerve problem. Can you afford to miss a brain tumor?

Optic nerve atrophy as manifested on fundus examination will generally take the form of optic disc pallor. There are certain situations, for instance, when pallor is altitudinal that a previous ischemic event would be strongly favored. This is not the case in the patient described above. Although nerve fiber layer thinning is altitudinal, the pallor is diffuse and the field defect if anything, suggests some respect for the vertical (not horizontal) meridian. However, in the end, as is true of most cases, it is impossible to make this call definitively simply based on the ophthalmoscopic appearance. It is wise, in the setting of available resources, to work up all patients without a known history or cause for optic atrophy to try and identify a treatable cause for the optic nerve problem. This work up should include a neuroimaging study. The neuroimaging study would preferably be a magnetic resonance image (MRI) scan with orbital views, including gadolinium enhancement and fat saturation,

which is personally reviewed by the examining physician, looking for a compressive lesion of the anterior visual pathway.

Lee et al. looked at just such a series of 91 patients with isolated, unexplained optic atrophy and identified compressive lesions in 18 (20%). Bilaterality, progression, and age under 50 were common in the group found to have tumors.(1) In older patients, one could make an argument that a conservative course of follow up would not be unreasonable even if such a lesion were identified. However, clearly, if a significant lesion at the skull base such as a meningioma or pituitary adenoma was identified in the patient with optic atrophy and visual field loss, then decompression through surgical resection and/or radiation therapy would be an important consideration.

The quality of the visual field defect would also potentially favor work up. In any patient who has a visual field defect that has any relationship to the vertical meridian, that is a defect that is denser temporally or nasally and seems to respect or not cross the vertical midline, the stakes are much higher for a compressive lesion and imaging is mandatory. This is even true in patients with unilateral optic atrophy or vision loss, as lesions of the chiasm and intracranial optic nerve can certainly cause either temporal or nasal field defects with optic disc atrophy in only one eye. An MR scan might also yield important information concerning concomitant white matter disease secondary to demyelinating illness and multiple sclerosis. We now recognize that the MRI is a very sensitive and important tool for diagnosing demyelinating disease. This would be an important cause for optic atrophy even in the absence of a known history of optic neuritis and therefore a possible diagnosis of multiple sclerosis could be suggested on an MRI in patients who are worked up for isolated optic atrophy. This would be less likely in a patient in this age group.

There is another subset of patients with optic atrophy that have been shown to have a high incidence of intracranial lesions. Patients that present with acute visual loss symptoms that suggest possible optic neuritis or ischemic optic neuropathy, but have optic atrophy present at the time of their initial presentation may harbor optic nerve meningiomas, parasellar meningiomas, pituitary adenomas, and intracranial sarcoidosis.(2)

Finally, a patient with isolated optic atrophy should also have a laboratory work up to identify treatable causes for optic neuropathies. This can be particularly important in situations where other aspects of the history suggest untreated infections or risk for nutritional deficiency. The important entities to consider here are syphilitic optic atrophy, which in the tertiary form can present with diffuse field loss and isolated optic atrophy, although there are generally other neurologic manifestations of this condition.

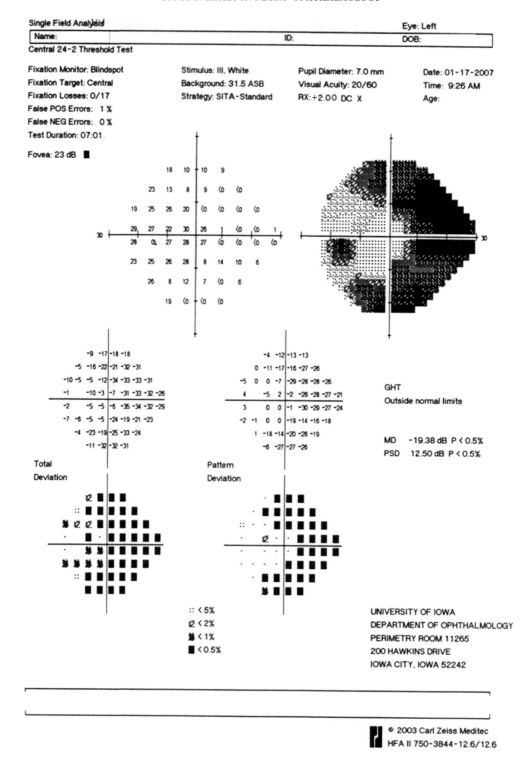

Figure 1.1 Humphrey Visual Field, left eye, showing significant visual field loss.

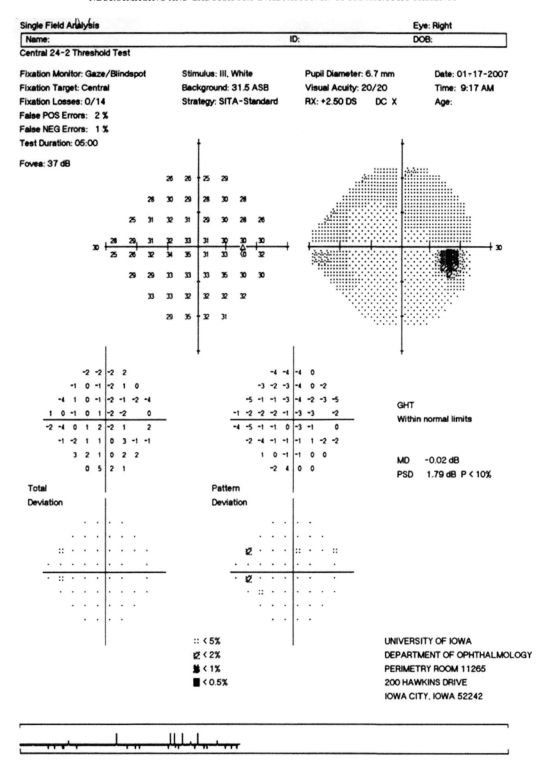

Figure 1.2 Essentially normal visual field, right eye.

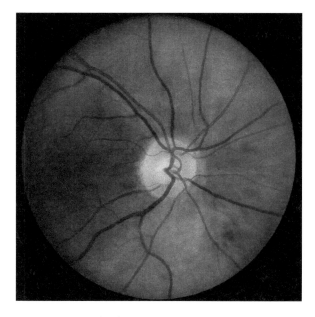

Figure 1.3 Normal right optic nerve.

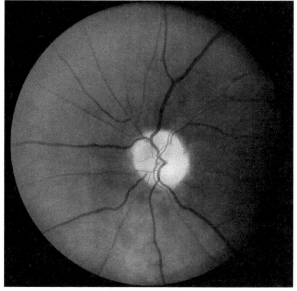

Figure 1.4 Left optic nerve showing disc pallor.

In other cases of bilateral vision loss with central scotoma and dyschromatopsia, nutritional causes for optic neuropathy should also be considered, particularly B12 deficiency.

In the end, any patient with an unexplained, newly recognized, neurologic deficit is owed the benefit of a workup. The clinician should depend on historical and examination clues to know how to direct this work up and without other identifiable cause for isolated optic atrophy the potential for a compressive lesion of anterior visual pathway must be excluded.

REFERENCES

1. Lee AG, Chau FY, Golnik KC, Kardon RH, Wall M. The diagnostic yield of the evaluation for isolated unexplained optic atrophy. Ophthalmology 2005; 112(5): 757–9.
2. Lee AG, Lin DJ, Kaufman M, Golnik KC, Vaphiades MS, Eggenberger E. Atypical features prompting neuroimaging in acute optic neuropathy in adults. Can J Ophthalmol 2000; 35(6): 325–30.

CON: A PATIENT WITH UNEXPLAINED ISOLATED OPTIC ATROPHY DOES NOT ALWAYS NEED NEUROIMAGING AND FURTHER LABORATORY EVALUATION

Karl Golnik

A complete ophthalmic examination including a comprehensive history will lead to an underlying diagnosis in 92% of cases of optic atrophy.(1) Ancillary studies such as neuroimaging and laboratory testing might be required to confirm the diagnosis but they are not necessary for every patient with optic atrophy.

The most common etiologies of optic neuropathy, nonarteritic anterior ischemic optic neuropathy (NAION), and optic neuritis are also the most common causes of optic atrophy.(1) Sudden, painless visual loss suggests a vascular etiology. Subacute painful visual loss favors inflammation and gradual visual loss may indicate a compressive or nutritional etiology. One caveat, the sudden discovery of chronic monocular visual loss may confound the history. Optic atrophy develops several months after damage and thus the patient who presents with acute or subacute visual loss (days to several weeks) and optic atrophy must have a more chronic process. If there has been no change over time, then one would consider static causes such as previous ischemia or trauma whereas progressive visual loss may indicate continued damage from compression or nutritional deficits. When considering ischemic optic neuropathy it is important to verify the presence of previous optic disc swelling.

Past medical history such as multiple sclerosis, severe vascular disease, sarcoidosis, or malignancy may suggest the cause of the optic atrophy. History of focal paresthesias or weakness may indicate demyelinating disease and shortness of breath and/or skin rash may occur with sarcoidosis. Gradual bilateral visual loss in other family members suggests possible dominant optic atrophy whereas a maternal family history suggests Leber's hereditary optic neuropathy. Finally, toxic exposures (methanol), contact with animals (cats, ticks), medications (ethambutol), and vitamin deficiencies (history of alcoholism) may direct diagnostic evaluation.

Clues in the ophthalmologic exam may aid in determining the underlying etiology of optic atrophy. Anterior segment exam may reveal evidence of previous trauma such as iris tears.

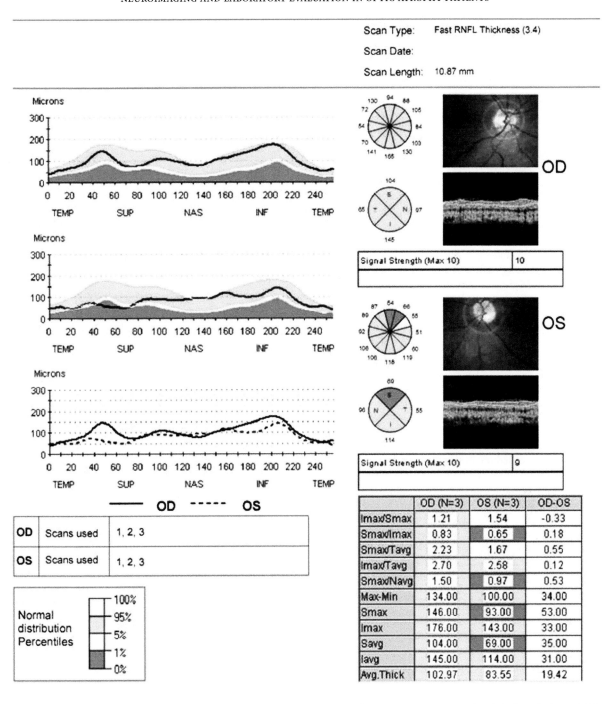

Figure 1.5 OCT of the RNFL showing nerve fiber layer loss on the left.

Additionally, the presence of active or previous inflammation such as keratitic precipitates or vitreous cell may point toward an infectious or inflammatory cause of optic atrophy such as sarcoid, syphilis, cat scratch disease, or Lyme disease. Hertel exophthalmometry may detect subtle proptosis and should be performed in every patient with optic atrophy. Computerized automated perimetry may detect specific patterns of visual loss helpful in the differential diagnosis. Central scotomas occur more commonly in nutritional, hereditary, or toxic optic neuropathies. Hemianopic field deficits suggest chiasmal or retrochiasmal damage.

The pattern of optic atrophy may be helpful. Diffuse optic atrophy and temporal optic atrophy are nonspecific, but an altitudinal pattern to the disc pallor is most often seen following the acute swelling in nonarteritic anterior ischemic optic neuropathy (NAION). Remember to confirm that the contralateral disc is small and congested (the disc-at-risk) when entertaining the diagnosis of NAION. Horizontal band (or "bow-tie") atrophy may be present with optic chiasmal or retrochiasmal pregeniculate lesions. Optociliary collateral vessels may become apparent when retinal venous outflow is compromised by an optic nerve sheath meningioma. Of course, one must exam the nerve with slit lamp biomicroscopy to obtain a good 3-dimensional view and to rule out subtle cupping that might occur in glaucoma.

Thus, laboratory testing such as angiotensin-converting enzyme, fluorescent Treponemal antibody (FTA-ABS), Lyme titer, and cat scratch titer (Bartonella henselae), Leber's hereditary optic neuropathy, or dominant optic atrophy (OPA1) may prove useful but only when history or examination has suggested the possibility of one of these diseases. Lab tests for infectious causes of optic atrophy can produce false positive results and are not useful without clinical correlation.(2, 3)

Neuroimaging is indicated in patients thought to have previous optic neuritis because of the association with multiple sclerosis. Any patient with optic atrophy and specific examination findings such as optociliary collateral vessels, proptosis, or visual field defects that respect the vertical midline should have neuroimaging.

Recently, we reported imaging results of 91 patients referred with unexplained, isolated, unilateral optic atrophy.(1) Twenty percent of these patients had compressive lesions demonstrated by magnetic resonance imaging with fat suppression and gadolinium administration. This study was done in two tertiary neuroophthalmology centers and thus may not be applicable to other patient care settings. Nevertheless, if there are no clues in the history or examination and no documentation of visual stability, then an MRI with gadolinium and fat-suppression should be obtained. If an MRI has been previously obtained, then I would review the films.

Unfortunately, a definite cause of the atrophy is not always discovered. If the history and appropriate evaluation do not produce a definite diagnosis, then repeat examination including automated perimetry should be done in 3 months to be sure there is no progressive loss of vision. If the exam is stable, repeat evaluations should occur on several occasions over the next 2 years to prove stability. If vision worsens during follow-up, repeat diagnostic testing is necessary.

REFERENCES

1. Lee A, Chau F, Golnik K, Kardon R, Wall M. The diagnostic yield of the evaluation for isolated unexplained optic atrophy. Ophthalmology 2005; 112(5): 757–9.
2. Sander A, Posselt M, Oberle K. Seroprevalence of antibodies to Bartonellae henselae in patients with cat scratch disease and in healthy controls: evaluation and comparison of two commercial serological tests. Clin Diag Lab Immunol 1998; 5(4): 486–90.
3. Bakken L, Callister S, Wand P. Interlaboratory comparison of test results for detection of Lyme disease in 516 participants in the Wisconsin state laboratory of hygeine/college of American pathologists proficiency testing program. J Clin Microbiol 1997; 35(3): 537–43.

SUMMARY

The decision whether to evaluate a patient with optic atrophy needs to include consideration of multiple factors including the absence (i.e., neurologically isolated) or presence (nonisolated) of other neurologic or systemic findings, the duration of the optic atrophy (i.e., chronic or subacute), the course (progressive or static), the laterality (unilateral or bilateral), the type of visual field defect (central scotoma, hemianopia), and the patient risk factors for a biologically plausible mechanism. The evaluation of optic atrophy could potentially include an extensive array of costly and potentially low yield tests (e.g., neuroimaging, laboratory testing, lumbar puncture, etc.). Patients with optic atrophy may require more or less evaluation depending on the comfort level of the clinician with the specific case and the pretest likelihood of a diagnosis. We recommend testing for patients in whom clinical uncertainty for the diagnosis is high and suggest directing the laboratory testing rather than a "shotgun" approach. It is important to involve the patient in the decision as well, as many patients cannot tolerate any uncertainty regarding even a remote possibility of a compressive lesion.

2 Should a young patient with a new diagnosis of optic neuritis have testing and treatment for multiple sclerosis (MS)?

A 24-year-old medical student in good general health began experiencing periocular pain OD 1 week ago. It was noticed to be worse with eye movement. Three days later she noticed decreased vision OD which became progressively worse for two additional days. On examination, her visual acuity is count fingers OD and 20/20 OS. There is a 1.2 log unit RAPD OD. Goldman perimetry shows a dense central scotoma OD (Figure 2.1 and 2.2). Ocular motility is full but slightly painful. The patient states that the pain is much less than the previous few days. There is no nystagmus. Anterior segment examination showed no uveitis. Dilated fundus exam was completely normal, with normal optic nerves and normal maculas OU (Figures

2.3 and 2.4). She denies any previous neurological symptoms including numbness, weakness, vertigo, or incontinence.

PRO: A YOUNG PATIENT WITH A NEW DIAGNOSIS OF OPTIC NEURITIS SHOULD HAVE TESTING AND TREATMENT FOR MULTIPLE SCLEROSIS IF IDENTIFIED

Fiona Costello

The evaluation and management of optic neuritis (ON) is not without controversy. There is a strong association between ON and multiple sclerosis (MS), which prompts the need for neuroimaging and raises questions regarding benefits to be derived

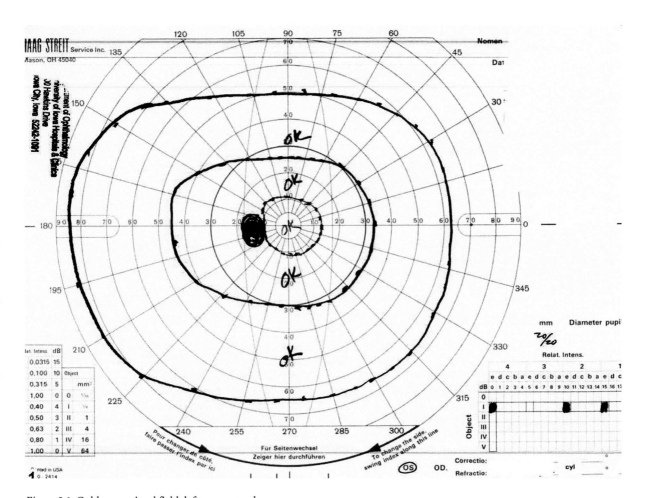

Figure 2.1 Goldmann visual field, left eye, normal.

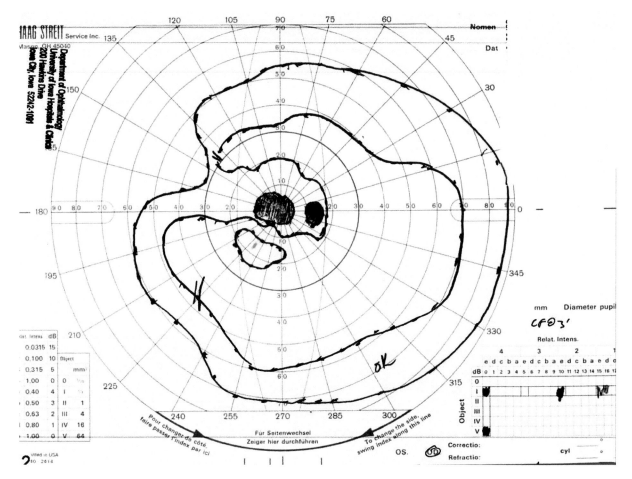

Figure 2.2 Goldmann visual field, right eye, large central scotoma.

from early initiation of disease-modifying therapy. Yet, not all ON patients harbor the same risk for future MS and potential MS-related disability. Therefore, efforts to follow, evaluate, and treat these patients must be tailored to meet the needs of the individual.

The Optic Neuritis Treatment Trial (ONTT) (1) demonstrated that the majority of ON patients share many of the features exemplified by the patient in the case example provided. More specifically, patients tend to be young (mean age 31.8 years) and female.(1, 2) Furthermore, ON patients often describe vision loss, which progresses over days to weeks; and 92% of patients report pain with eye movement. Unilateral ON is associated with a relative afferent pupil defect in the affected eye, and the pattern of visual field tends to respect the topography of the retinal nerve fiber layer. Visual recovery from ON tends to occur 4–6 weeks after onset, and improvement may continue for up to 1 year. (3) ON remains first and foremost a clinical diagnosis and

the typical clinical syndrome DOES NOT generally require additional investigations including blood work, electrophysiology, or orbital imaging.

The real impetus for investigations in patients with suspected ON stems from its strong association with MS. Approximately 20% of MS patients will present with ON as their first demyelinating event. The ONTT showed that the baseline magnetic resonance imaging (MRI) scan was the most potent predictor for the development of future MS. At 10 years, ON patients with 1 or more white matter lesions on the baseline MRI study had a 56% risk of developing clinically definite MS (CDMS), whereas ON patients with no lesions had a 22% risk of CDMS at 10 years.(4) Therefore MR imaging is integral to the evaluation of ON patients.

Additional studies including visual-evoked potentials (VEP), cerebrospinal fluid (CSF) analysis, and optical coherence tomography (OCT) can also provide valuable information in the evaluation of ON patients. VEP abnormalities are

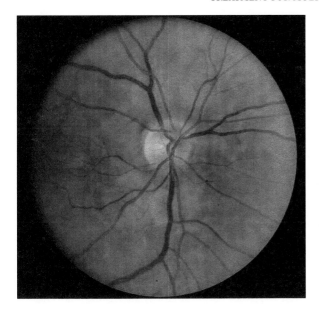

Figure 2.3 Normal optic nerve, right eye.

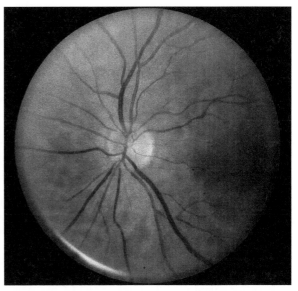

Figure 2.4 Normal optic nerve, left eye.

common in MS, and testing can unveil clinically occult lesions, which provide evidence of dissemination in ON patients undergoing evaluation for possible MS.(2) Cerebrospinal fluid (CSF) irregularities, including abnormal intrathecal IgG synthesis (defined as 2 or more oligoclonal bands in the CSF without corresponding bands in the serum), occur in 60–70% of patients with ON and other clinically isolated syndromes (CIS). A long-term Swedish study of 86 patients with acute monosymptomatic unilateral ON showed that patients with signs of inflammation in the CSF (raised cell count, oligoclonal bands, or both) had a 49% risk of future MS as compared to patients with no CSF abnormalities (23%).(5, 6) OCT is a novel imaging technique, which can be used to detect and quantify the effects of retrograde axonal degeneration due to ON, by measuring thinning in the retinal nerve fiber layer (RNFL) of the eye. OCT-measured RNFL values are diminished among ON and MS patients.(7–17) Furthermore, reduced RNFL values in ON and MS patients have been shown to correlate with: diminished visual and neurological function,(7–11, 14, 16, 17) reduced optic nerve magnetization transfer ratios, (13) MRI-measured optic nerve and brain atrophy, (12, 15) and decreased cerebral brain matter volumes.(14) Recent publications have highlighted the potential role for OCT-measured RNFL values as a candidate biomarker for axon loss in the study of ON and MS.(9, 14, 16, 18)

There is also a role for more detailed investigations when atypical clinical features are encountered. Lack of clinical improvement after presumed ON can implicate an underlying compressive mass, such as a nerve sheath tumor or suprasellar mass. In this context, patients may report "sudden onset" monocular vision loss, which represents acute awareness, rather than acute onset of symptoms. Similarly, the presence of optic disc pallor is not consistent with the diagnosis of acute ON, and can herald an underlying compressive optic neuropathy. In both scenarios, neuroimaging can reveal a culprit lesion. If abundant vitreous cells, macular edema, or florid optic disc edema is observed then infectious neuroretinitis may be a more tenable diagnosis than ON. Careful clinical follow up will often disclose the development of a "macular star" in these cases. Older individuals, with vascular risk factors may occasionally present with pain and vision loss due to anterior ischemic optic neuropathy. In such cases, the diagnosis can usually be determined by a careful clinical history, detailed examination (which generally demonstrates significant optic disc swelling), and the observation of less complete visual recovery. In addition, these patients often have a specific morphological appearance, with a small or absent physiological cup to suggest the diagnosis. Posterior ischemic optic neuropathy cases may be more challenging to distinguish from optic neuritis, as optic disc swelling is not apparent at the time of clinical presentation. Again, a detailed clinical history and examination should disclose this diagnosis, which is often one of exclusion. Clinical manifestations of other systemic diseases such as rash, joint pain, alopecia, hematological abnormalities, renal failure, and/or opportunistic infections should be thoroughly investigated for other potential etiologies.

Neuromyelits Optic (NMO) or Devic's Syndrome is a severe inflammatory process of the optic nerves and spinal cord, often associated with poor clinical recovery. Within 5 years, 50% of patients afflicted with NMO have irreversible vision loss in one eye or can no longer ambulate independently. In addition to optic nerve involvement, typical features of NMO include: episodic myelitis (with spinal lesions that extend 3 or more spinal segments), absence of clinical manifestations of brain involvement, and absence of typical brain MRI lesions.(18–20) The early recognition of this clinical syndrome and more specifically the ON manifestations that may herald this diagnosis are important, as the treatment for NMO differs from that of MS. Immunosuppressive therapies including plasma exchange therapy, azathioprine, rituximab, and corticosteroids are more effective therapies than the immunomodulating alternatives (interferons and glatiramer acetate), which are more frequently implemented in the management of MS. Ideally, if treated early, some of the more disabling features of the complete clinical Devic syndrome might be prevented with early initiation of immunosuppressive therapy. I consider the diagnosis of NMO in patients with poor visual recovery after ON, symptoms of myelitis, and negative cranial MR imaging. For this reason, I often include a cervical spine study in the baseline MRI study to check for occult, extensive spinal lesions. Recently, Lennon and colleagues described a putative marker, NMO-IgG autoantibody (sensitivity 73% and specificity 91%), which binds at or near the blood–brain barrier, and distinguishes NMO from MS.(20) For patients considered to be at high risk for NMO, testing for the NMO-IgG antibody might help expedite the necessary treatment regimen for this distinct clinical syndrome.

The role for therapy in ON is not to expedite recovery of optic nerve function, which tends to be good, but rather to impact the future risk of MS. Three studies have both addressed the role of interferon therapy for acute monosymptomatic ON, and the future development of MS. The first of these was the Controlled High Risk Subjects Avonex Multiple Sclerosis Prevention Study (CHAMPS) (21), in which 383 CIS patients (with prior ON; brainstem or cerebellar syndrome; or an incomplete transverse myelitis) were enrolled into a randomized, placebo-controlled trial if they had 2 or more clinically silent lesions on a cranial MRI scan. Fifty percent (192 patients) of the CIS patients enrolled in this study had ON. After initial treatment with high dose intravenous methylprednisolone, half the patients received weekly interferon beta-1a (30 µg once per week), and half received placebo. The primary end point was the development of CDMS, and the secondary end point was the brain MRI. This study demonstrated a significantly lower rate (44%) of development of CDMS among the treatment group (rate ratio 0.56; 95% CI 0.38–0.81; $p = 0.002$), and a relative reduction of new lesions in the cranial MRI scans among patients treated with interferon *versus* the placebo group. A second study, Early Treatment of MS (ETOMS), (22) enrolled 308 patients, with 4 asymptomatic white matter lesions (or 3 lesions if one enhanced with gadolinium) on the cranial MRI scan at presentation. Half the patients received subcutaneous interferon beta-1a (22 µg once a week), and half received placebo. After 2 years, the odds ratio for the development of CDMS was 0.61 (95% CI 0.37–0.99; $p = 0.045$) in the treatment group *versus* the control group. More specifically, 45% of the placebo group developed CDMS after 2 years as compared to 34% of treated

patients. During the treatment study period, the MRI activity and burden of disease measured by MRI were significantly reduced in the treatment group. In the third and most recent study which looked at the role of disease modifying therapy in CIS, the Betaferon in Newly Emerging Multiple Sclerosis for Initial Treatment (BENEFIT) (23) trial, CIS patients with at least 2 clinically silent brain MRI lesions were randomized to receive interferon beta-1b 250 mcg subcutaneously on alternate days or placebo until CDMS was diagnosed or the study period of 24 months was reached. Overall interferon beta-1b delayed the time to diagnosis of CDMS and McDonald criteria defined MS.(23) Hence, there is evidence that disease modifying therapy may be indicated in patients with acute monosymptomatic ON who are deemed to be at high risk on the basis of their MRI findings, to prevent or delay the development of CDMS.

The question of whether to initiate disease-modifying therapy after isolated, monosymptomatic ON remains a controversial one. Recently, the pros and cons of this debate were presented in the Archives of Neurology.(24, 25) Frohman (24) and colleagues put forth an elegant argument favoring early initiation of disease modifying drugs for patients with MS or a CIS, and cited a number of reasons to support early treatment in CIS and MS patients. These issues outlined by these authors apply well to ON as CIS patients, and are as follows:

- MS is a disabling illness: Many patients with ON will develop MS, and the majority of MS patients develop significant disability over time. It is impossible to predict a benign course and to forgo treatment based on this assumption will result in the accumulation of irreversible disability for some patients.
- Irreversible axonal loss occurs early in MS: Pathological and radiological studies show that irreversible axonal injury occurs early in MS patients, which might not be detected in clinical observation. The authors also point out that current therapies are not reparative, but are preventive in action.
- Approved therapies are available: The Food and Drug Administration (FDA) has approved medications that work best early in the course of MS, even at the time of a CIS, and less effective in progressive phases of the disease.
- Potentially dire long-term consequences: Delay in therapy has been associated with a greater burden of disease on MRI and in the number of patients with progression of disability.

Optic Neuritis is an important clinical entity, which carries with it an association and future risk of multiple sclerosis (MS). Cranial MR imaging is integral to the evaluation of ON, because it represents the most potent predictor for the later development of MS. Additional investigations with VEP, CSF and OCT studies can enhance the evaluation of ON patients, and help to exclude potential mimics. The question of whether disease-modifying therapy should be initiated after ON as a CIS is a controversial topic, and factors specific to the patient should be taken into consideration before weighing in favor or against this therapeutic option. A significant proportion of ON patients will go on to develop future MS, and early initiation of therapy may delay this diagnosis. In addition, disease-modifying therapies may also reduce the disabling effects of MS among patients. There are currently approved therapies available for patients with ON in whom a baseline MRI scan reveals the presence of white matter lesions, which predict a greater risk of future MS. Therefore, I advocate using ancillary testing to try and identify "high risk" ON patients who are likely to develop MS, and treat accordingly.

REFERENCES

1. Beck RW, Cleary PA, Anderson MM. A randomized controlled trial of corticosteroids in the treatment of acute optic neuritis. N Engl J Med 1992; 326: 581–88.
2. Hickman SJ, Dalton CM, Miller DH. Management of acute optic neuritis. Lancet 2002; 360: 1953–62.
3. Optic Neuritis Study Group. Visual function 5 years after optic neuritis. Arch Ophthalmol 1997; 115: 1545–52.
4. Beck RW, Trobe JD, Moke PS et al. High- and low-risk profiles for the development of multiple sclerosis within 10 years after optic neuritis: experience of the Optic Neuritis Treatment Trial. Arch Ophthalmol 2003; 121(7): 944–9.
5. Miller D, Barkhof F, Montalban X et al. Clinically isolated syndromes suggestive of multiple sclerosis, part 1: natural history, pathogenesis, diagnosis and prognosis. Lancet Neurol 2005; 4: 281–88.
6. Nilsson P, Larsson EM, Maly-Sundgren P et al. Predicting the outcome of optic neuritis: evaluation of risk factors after 30 years of follow up. J Neurol 2005; 252(4): 396–402
7. Parisi V, Manni G, Spadaro M et al. Correlation between morphological and functional retinal impairment in multiple sclerosis patients. Invest Ophthalmol Vis Sci 1999; 40(11): 2520–7.
8. Trip SA, Schlottmann PG, Jones SJ et al. Retinal nerve fiber layer axonal loss and visual dysfunction in optic neuritis. Ann Neurol 2005; 58(3): 383–91.
9. Fisher JB, Jacobs DA, Markowitz CE et al. Relation of visual function to retinal nerve fiber layer thickness in multiple sclerosis. Ophthalmology. 2006; 113(2): 324–32.
10. Noval S, Contreras I, Rebolleda G, Munoz-Negrete FJ. Optical coherence tomography versus automated perimetry for follow-up of optic neuritis. Acta Ophthalmol Scand 2006; 84(6): 790–4.
11. Costello F, Coupland S, Hodge W et al. Quantifying axonal loss after optic neuritis with optical coherence tomography. Ann Neurol 2006; 59(6): 963–9.
12. Trip SA, Schlottmann PG, Jones SJ et al. Optic nerve atrophy and retinal nerve fiber layer thinning following optic neuritis: evidence that axonal loss is a substrate of MRI-detected atrophy. Neuroimage 2006; 31(1): 286–93.
13. Trip SA, Schlottmann PG, Jones SJ et al. Optic nerve magnetization transfer imaging and measures of axonal loss and demyelination in optic neuritis. Mult Scler 2007; 13(7): 875–9.
14. Sepulcre J, Murie-Fernandez M, Salinas-Alaman A et al. Diagnostic accuracy of retinal abnormalities in predicting disease activity in MS. Neurology 2007; 68(18): 1488–94.
15. Gordon-Lipkin E, Chodkowski B, Reich DS et al. Retinal nerve fiber layer associated with brain atrophy in multiple sclerosis. Neurology 2007; 69(16): 1603–9.
16. Pulicken M, Gordon-Lipkin E, Balcer LJ et al. Optical coherence tomography and disease subtype in multiple sclerosis. Neurology 2007; 69(22): 2085–92.
17. Henderson APD, Trip SA, Schlottmann PG et al. An investigation of the retinal nerve fiber layer in progressive multiple sclerosis using optical coherence tomography. Brain 2008; 131: 277–87.
18. Lennon VA, Wingerchuk DM, Kryzer TJ et al. A serum autoantibody marker of neuromyelitis optica: distinction from multiple sclerosis. Lancet 2004; 354: 2106–12.
19. Cree BAC, Goodin DS, Hauser SL. Neuromyelitis Optica. Semin Neurol 2002; 22: 105–122.
20. Wingerchuk DM, Hogancamp WF, O'Brien PC, Weinshenker BG. The clinical course of neuromyelitis optica (Devic's syndrome). Neurology 1999; 53: 1107–14.
21. Jacobs LD, Beck RW, Simon JH et al. Intramuscular interferon beta-1a therapy initiated during a first demyelinating event in multiple sclerosis. CHAMPS Study Group. N Engl J Med 2000; 343(13): 898–904.
22. Comi G, Filippi M, Barkof F et al. Effect of early interferon treatment on conversion to definite multiple sclerosis: a randomised study. Lancet 2001; 357(9268): 1576–82.
23. Kappos L, Polman CH, Freedman MS et al. Treatment with interferon beta-1b delays conversion to clinically definite and McDonald MS in patients with clinically isolated syndromes. Neurology 2006; 67(7): 1242–9.
24. Frohman EM, Havrdova E, Lublin F et al. Most patients with multiple sclerosis or a clinically isolated demyelinating syndrome should be treated at the time of diagnosis. Arch Neurol 2006; 63(4): 614–19.
25. Pittock SJ, Weinshenker, Noseworthy JH et al. Not every patient with multiple sclerosis should be treated at time of diagnosis. Arch Neurol 2006; 63(4): 604–11.

CON: A YOUNG PATIENT WITH A NEW DIAGNOSIS OF OPTIC NEURITIS DOES NOT ALWAYS REQUIRE TESTING AND TREATMENT FOR MS

Michael S Lee

TESTING

The Optic Neuritis Treatment Trial (1) collected valuable data on a large cohort of patients with optic neuritis in a standardized fashion. Each of the 448 patients underwent a brain MRI, chest X-ray, and serologic testing (blood glucose, ANA, and FTA-ABS). Although optional, 141 patients underwent lumbar puncture. Of all the patients with optic neuritis, only one (0.2%) developed a connective tissue disease. Positive syphilis testing occurred in 6 (1.3%) patients but repeat testing yielded negative results. None of the chest X-ray images demonstrated significant findings. Meanwhile, lumbar punctures revealed modest elevation of protein in 10% and white cell count in 36%, consistent with mild inflammation. However, cerebrospinal fluid (CSF) analysis did not change the diagnosis or yield another disorder in any patient.

The patient described here has a classic story for isolated, unilateral optic neuritis and further testing is unlikely to affect the *diagnosis*. The patient fits the right demographic and her symptoms and signs all comport with the diagnosis. Patients with optic neuritis are generally between the ages of 20–40 years of age. The pain often precedes the visual loss and lasts < 10 days. It generally worsens with touching or moving the eye. The most common visual field defect is a central scotoma and the optic nerve appears normal in 2/3 of cases. If the patient demonstrated atypical features, then I would consider further evaluation.

Since neuroimaging of the brain affects the future risk for the development of multiple sclerosis, I think it is reasonable in all cases for prognostication. Approximately 2/3 of the patients in the ONTT completed 15-year follow up.(2) Overall, 50% received a diagnosis of clinically definite multiple sclerosis (CDMS). Further analysis showed that 25% of patients with a normal baseline brain MRI developed CDMS. This is compared to nearly 3/4 of patients with at least one typical white matter lesion on brain MRI developed CDMS. There was no significant difference in conversion rates between patients with one white matter lesion and more than one lesion. I do not believe that a lumbar puncture is necessary in typical cases. The presence of oligoclonal banding in CSF predicted the future development of CDMS in the ONTT, but this predictive capacity was not independent of the MRI.

TREATMENT

Visual function recovers more quickly among patients receiving intravenous (IV) corticosteroids compared to placebo, but corticosteroids do not affect final visual acuity or field. Patients who receive IV corticosteroids with abnormal neuroimaging may reduce the risk of CDMS for up to 2 years. It may be reasonable to consider intravenous corticosteroids followed by a 2-week oral taper if the patient has an abnormal brain MRI. Interestingly, oral corticosteroids alone increase the risk of recurrent optic neuritis but not CDMS. I would not recommend oral corticosteroids.

Interferon beta therapy reduces the conversion to CDMS compared to placebo among patients who have both optic neuritis and two or more characteristic white matter lesions. It is not a panacea and approximately 1/3 of patients who start therapy still develop CDMS in the first 3–5 years. Immunomodulatory therapy has never been studied among patients with < two lesions, so I would not advocate interferon beta therapy in this group. In the ONTT, 25% of patients with an abnormal brain MRI did not develop CDMS at 15-year follow up. Therefore, initiating interferon therapy in all patients with an abnormal brain MRI may expose a group of patients to unnecessary life-long therapy. Even among patients who develop CDMS, severity of disease progression is highly variable and unpredictable. Up to 1/3 of patients with relapsing remitting CDMS have a relatively benign prognosis at 10-year follow up and may not require disease modifying therapy. These patients may enjoy limited benefit from therapy and suffer from unwanted side effects and cost.

Common side effects of self injectable MS drugs include injection site reactions, flu like symptoms, depression, and chest pain. Approximately 10–20% of patients who have MS discontinue use of these drugs because of adverse side effects. Interferon beta therapy costs approximately $15,000–24,000 per year, which may be prohibitive for many patients and a significant strain on the health care system.

REFERENCES

1. Optic Neuritis Study Group. The clinical profile of optic neuritis: experience of the Optic Neuritis Treatment Trial. Arch Ophthalmol 1991; 109: 1673–78.
2. Optic Neuritis Study Group. Multiple sclerosis risk after optic neuritis: final Optic Neuritis Treatment Trial follow-up. Arch Neurol 2008; 65: 727–32.

SUMMARY

Patients with new optic neuritis may or may not ultimately develop multiple sclerosis. Factors that might predict a future diagnosis of MS (e.g., family history of MS, MRI showing demyelinating white matter lesions, prior attacks, or subjective neurologic symptoms) should be considered in the decision making. Just as importantly however factors that might predict a lower risk for future MS should be considered as well (e.g., male patient with no light perception vision, lack of pain, macular exudate, or normal MRI). The important part is that the patient be involved in the decision making. If the diagnosing ophthalmologist is unwilling or unable to have this discussion with the patient then consultation with a neurologist or neuroophthalmologist should be considered regarding the options for neuroimaging of the head or spine, additional laboratory testing for MS mimics, a lumbar puncture for oligoclonal bands and other markers of demyelination, and possible MS treatment (e.g., immunomodulatory agents). Although a case can be made for minimal or no work up or treatment for clinically isolated syndromes the patient should be allowed to participate in the discussion and decision and provided with sufficient information to make an informed choice.

3 Should a patient with optic disc edema with a macular star figure (neuroretinitis) have lab testing and treatment?

A 24-year-old female presents to the local ophthalmologist with a complaint of decreased vision in the right eye. She noticed it gradually over a 3-day period occurring 2 weeks before presentation, and it has remained the same since. She has no other neurologic symptoms and no other medical history. She specifically denies any prior joint pain. She lives in the upper Midwest, where she has 2 dogs and 1 cat. She has had no tick bites, cat-scratch, or travel history. She denies any history of sexually transmitted disease. On examination the visual is 20/200 OD and 20/15 OS. There is a 1.2 log unit RAPD OD. Slit lamp examination reveals no evidence of uveitis. Visual fields show a cecocentral scotoma OD, and a full field OS (Figures 3.1 and 3.2). She has optic disc edema with a macular star pattern of exudate OD and a normal disc and macula OS (Figures 3.3 and 3.4). OCT of the RNFL (Figure 3.5) and macula (Figure 3.6) show the disc edema, with extension of the fluid under the fovea OD.

PRO: TEST FOR CAT SCRATCH, LYME, SYPHYLLIS, TUBERCULOSIS (TB) AND TREAT EMPIRICALLY FOR CAT SCRATCH FEVER

Karl Golnik

Neuroretinitis is defined as the presence of optic disc swelling and macular exudate. The exudate typically takes the form of

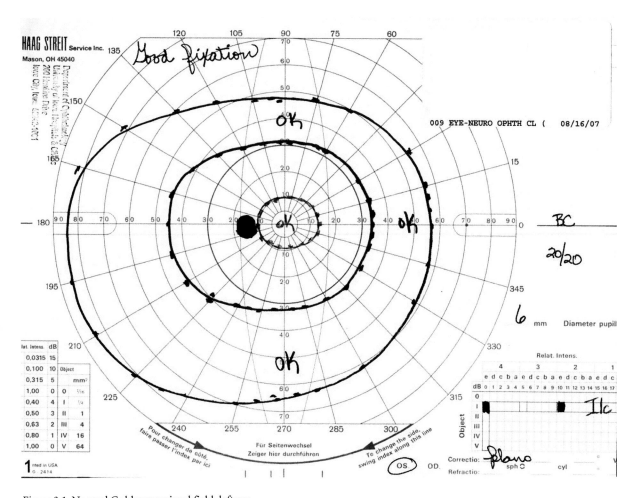

Figure 3.1 Normal Goldmann visual field, left eye.

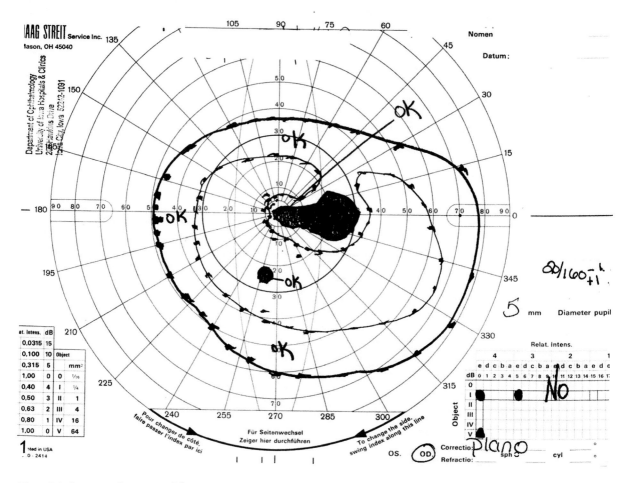

Figure 3.2 Cecocentral scotoma, right eye.

a star or partial star as the exudate accumulates in the radially oriented nerve fiber layer of Henle. Patients typically experience subacute loss of central vision over several days. If the patient presents within the first few days of visual loss the nerve may appear swollen and no exudate may be apparent. Usually the macula will appear thickened and presumably the exudate has not had a chance to accumulate that quickly. Within a week or so the exudates coalesce to form the star. When a patient is seen early in the course before star formation, typical optic neuritis may be misdiagnosed, multiple sclerosis unnecessarily discussed, and MRI ordered. Thus, it is crucial to scrutinize the macula with slit lamp biomicroscopy and consider macular OCT if there is any question of macular abnormality. Traditionally, neuroretinitis is felt to be a self-limited condition with good recovery of vision. However, visual recovery may not occur and certainly may not return to completely normal.(1, 2)

A variety of conditions have been reported to cause neuroretinitis. These include postviral, cat scratch disease (Bartonella henselae), Lyme, syphilis, tuberculosis, toxoplasmosis, and sarcoidosis. Hypertension, papilledema, and nonarteritic anterior ischemic optic neuropathy may produce the picture of neuroretinitis but are not considered true neuroretinitis because it is felt to be an inflammatory condition. Hypertension and papilledema should produce bilateral findings and one should check blood pressure and ask about symptoms of elevated intracranial pressure (headache, tinnitus, nausea) in this circumstance. Lyme disease, syphilis, toxoplasmosis, tuberculosis, and sarcoidosis are thought to be fairly rare causes of neuroretinitis. Patients should be asked about immune status, endemic area exposure, sexual history, skin/genital rash or lesions, tuberculosis exposure, and history of uveitis. The entity of idiopathic retinitis, vasculitis, aneurysms, and neuroretinitis (IRVAN) must be also be considered but the associated retinal findings should differentiate this condition from typical neuroretinitis.(3)

Cat scratch disease is by far the most common identifiable cause of neuroretinitis.(1) Bartonella henselae is a small

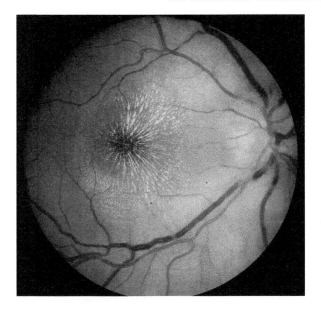

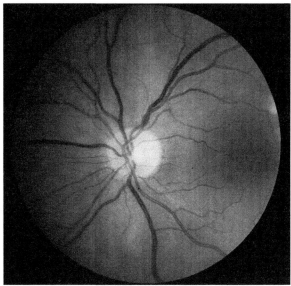

Figure 3.3 Fundus photograph, right eye, showing classic macular star.

Figure 3.4 Fundus photograph left eye, normal.

gram-negative rod that has been shown to be the cause of cat scratch disease. Patients usually develop a mild to moderately severe flu-like illness associated with regional lymphadenopathy. Common systemic symptoms include fever, headache, anorexia, nausea, vomiting, and sore throat. Ocular involvement occurs in 5%–10% of patients with cat scratch disease. Other less common manifestations of cat scratch disease include encephalitis (1%–2%), osteomyelitis (less than 1%), and hepatosplenic disease (less than 1%).(4) Ocular manifestations include neuroretinitis, Parinaud's oculoglandular syndrome, multifocal retinitis/choroiditis, retinal white spots, and retinal vascular occlusion. The most practical means of laboratory diagnosis is serology for Bartonella henselae antibodies.

Treatment of cat scratch disease is somewhat controversial because of its historically benign course. At the time of this manuscript a variety of antibiotics have been shown to be effective *in vitro*. Doxycycline and erythromycin have also been shown to produce good results in treating *Bartonella henselae* infection in immunocompromised patients. Margileth and associates reviewed 268 immunocompetent patients with systemic cat scratch disease and found untreated patients and patients treated with antibiotics subsequently thought not to be effective had a mean duration of illness of 14.5 weeks *versus* 2.8 week duration of illness in the group treated with antibiotics thought to be efficacious (rifampin, ciprofloxacin, gentamicin, trimethoprim-sulfamethoxazole).(5) There have been no randomized trials regarding treatment of neuroretinitis caused by *B. henselae* infection. All reported treated cases are anecdotal in nature. Reed and associates treated 7 patients with oral

doxycycline (100 mg BID) and rifampin (300 mg BID) for 4–6 weeks. They compared these patients to historical reports and felt that the treatment shortened the course of disease and hastened visual recovery.(2)

Thus, in the patient with neuroretinitis testing for Lyme, syphilis, toxoplasmosis, sarcoidosis, and tuberculosis should be considered depending on prevalence in one's geographic area and whether any of the risk factors listed above are present. Serologic testing for *B. henselae* should always be obtained because it is frequently the cause of the neuroretinitis. Furthermore, it is important to note that not all patients who test positive for cat scratch disease have had or remembers having been scratched by a cat and sometimes they remember only in retrospect.

Treatment for *B. henselae* is more controversial. However, treatment should be considered for the following reasons:

1. The recommended antibiotics are fairly benign.
2. Treatment has been shown to be efficacious in immunocompromised patients with systemic disease.
3. Retrospective studies show that duration of systemic illness may be less if treated.
4. Not every patient has good recovery of vision and they want to feel that everything that can be done has been done.

Thus, I discuss empiric treatment with every patient pending lab results and when asked what I would do if it were my eye, I tell them I would take the medicine.

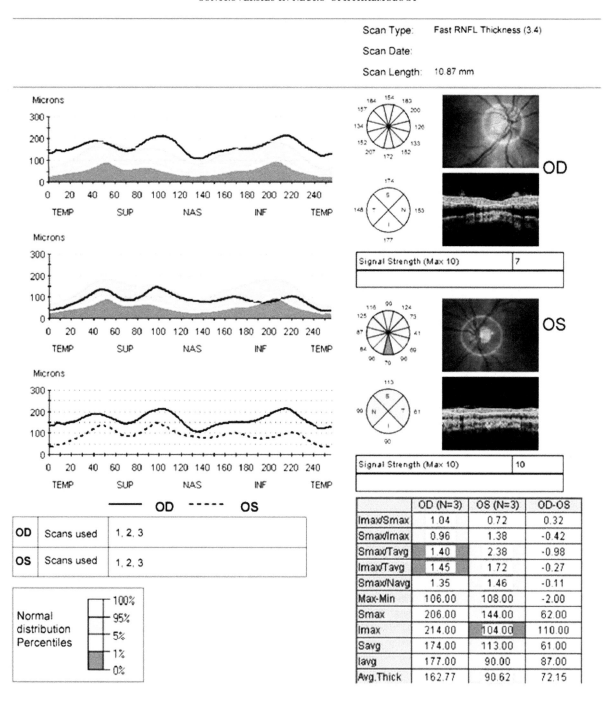

Figure 3.5 OCT showing the marked thickening of the RNFL on the right.

Scan Type: 7mm OD OD

Scan Date:

Scan Length: 7 0 mm

OCT Image

Fundus Image

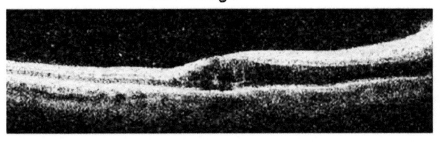

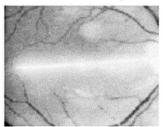

Signal Strength (Max 10) | 6

Scanned Image

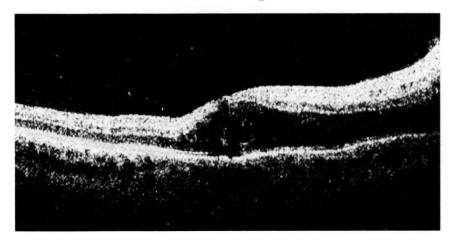

Figure 3.6 Line scan through the fovea showing macular edema associated with the disc edema.

REFERENCES

1. Suhler EB, Lauer AK, Rosenbaum JT. Prevalence of serologic evidence of cat-scratch disease in patients with neuroretinitis. Ophthalmology 2000; 107: 871–6.
2. Reed JB, Scales DK, Wong MT et al. Bartonella henselae neuroretinitis in cat scratch disease: diagnosis, management and sequelae. Ophthalmology 1998; 105(3): 459–66.
3. Samuel MA, Equi RA, Chang TS et al. Idiopathic retinitis, vasculitis, aneurysms and neuroretinitis (IRVAN). Ophthalmology 2007; 114: 1526–9.
4. Cunningham ET, Koehler JT. Ocular Bartonellosis. Am J Ophthalmol 2000; 130: 340–9.
5. Margileth AM. Antibiotic therapy for cat-scratch disease: clinical study of therapeutic outcome in 268 patients and a review of the literature. Pediatr Infect Dis J 1992; 11(6): 474–8.

CON: DO NOT TEST FOR CAT SCRATCH, LYME, SYPHYLLIS, TUBERCULOSIS (TB) AND DO NOT TREAT FOR CAT SCRATCH FEVER

Eric Eggenberger

Neuroretinitis is a distinct clinical neuroophthalmic presentation. In addition to an anterior optic neuropathy, leakage from incompetent retinal vessels leads to a macular star figure. The unique appearance serves to distinguish this clinical syndrome from more common anterior optic neuropathies such as optic neuritis and anterior ischemic optic neuropathy. There is an extensive differential diagnosis for neuroretinitis (Table), and a definitive identifiable cause confirmed in the minority. Accordingly, although a long list of labs may be considered, this is often not useful or treatment-altering.

Idiopathic neuroretinitis cases generally behave in a similar fashion to cases in which a lab abnormality points to a specific origin. An exhaustive series of lab tests significantly adds to the expense of managing such cases despite the high rate of negative findings. Furthermore, false positive and negative results occur more commonly when the clinician uses such widespread "shot gun" laboratory approaches, and these results can push the clinician down unnecessary and costly roads.

In addition, there are no studies demonstrating the value of therapy for the most common identifiable cause of neuroretinitis, Bartonella henselae. Thus, even in cases where antibodies to this agent are identified, the best management remains unknown, with many clinicians following a conservative approach. The exact risks, benefits, and side effects of a treatment course of steroids and antibiotics also remain unknown. In a review of 202 cases of cat scratch disease, antibiotics were associated with varying degrees of effectiveness, and the author recommended conservative, symptomatic treatment for the majority of patients with mild or moderate disease. The potential side effects of unproven treatments must also be kept in mind by the treating clinician, especially in children.

In conclusion, there are no large, masked trials to assist in clinical management of neuroretinitis. Although testing and therapy are reasonable positions when managing neuroretinitis, this approach may be costly, unnecessary in most, often unrevealing, and not mandatory or backed by evidence-based guidelines. Individual historical and examination features remain the most useful guideposts in management of this clinical condition.

Table Differential Diagnosis of Neuroretinitis.

1. Infectious
2. Bacterial
Bartonella henselae
Tuberculosis
Lyme
Syphilis
Leptospirosis
3. Viral
CMV
Herpes
EBV
4. Parasite
Toxoplasmosis
Toxocara
5. Inflammatory
sarcoidosis
6. Ischemic

EBV = Epstein Barr Virus; CMV = Cytomegalovirus.

BIBLIOGRAPHY

Wals KT, Ansari H, Kiss S et al. Simultaneous Occurrence of Neuroretinitis and Optic Perineuritis in a Single Eye. J Neuro-Ophthalmol 2003; 23(1): 42–27.

Margileth AM. Antibiotic therapy for cat-scratch disease: clinical study of therapeutic outcome in 268 patients and a review of the literature. Pediatr Infect Dis J 1992; 11(6): 474–8.

SUMMARY

The presence of optic disc edema and a macular star figure (ODEMS) typically is the herald for infectious "neuroretinitis". Although many authors have advocated for testing for treatable etiologies like cat scratch disease, Lyme disease, syphilis, or tuberculosis (TB), most cases are self-limited. Treatment with antibiotics empirically for cat scratch disease has not been proven to be efficacious but patients could be offered the option of treatment. This is especially reasonable considering the typically limited side-effects of treatment. Although the yield for testing for alternative etiologies for neuroretinitis other than cat scratch disease is low, it may be reasonable to pursue additional tests depending upon the pre-test likelihood of disease in a specific patient.

4 Should a vasculopathic patient with nonarteritic anterior ischemic optic neuropathy have any testing?

A 62-year-old man noted blurry vision in his lower left visual field on awakening 2 days ago. The visual symptoms are stable since onset. His past medical history is positive for hypertension treated with a beta-blocking agent daily, and also for hypercholesterolemia under control with diet. His last appointment with his primary care doctor was 1 year ago. On examination, his vision is 20/20 OD and 20/25 OS. There is a 0.9 log unit RAPD OS. There is an inferior altitudinal visual field defect OS and a full field OD (Figures 4.1 and 4.2). The optic nerve OD has a small cup/disc ratio and there is optic nerve swelling OS (Figures 4.3 and 4.4). He denies any symptoms of temporal arteritis. His erythrocyte sedimentation rate and C-reactive protein are normal.

PRO: TEST FOR BLOOD PRESSURE (NOCTURNAL HYPOTENSION, 24 HOUR BLOOD PRESSURE MEASUREMENTS), SLEEP APNEA, BLOOD SUGAR, CHOLESTEROL, NO SMOKING, ASPIRIN PER DAY

Karl Golnik

Patients with nonarteritic anterior ischemic optic neuropathy (NAION) typically are > 50 years of age and present with painless, sudden visual loss. Vision improves (3 or more lines) in about 43% of patients over 6 months.(1) Reported risk factors and associated conditions include age, hypertension, nocturnal

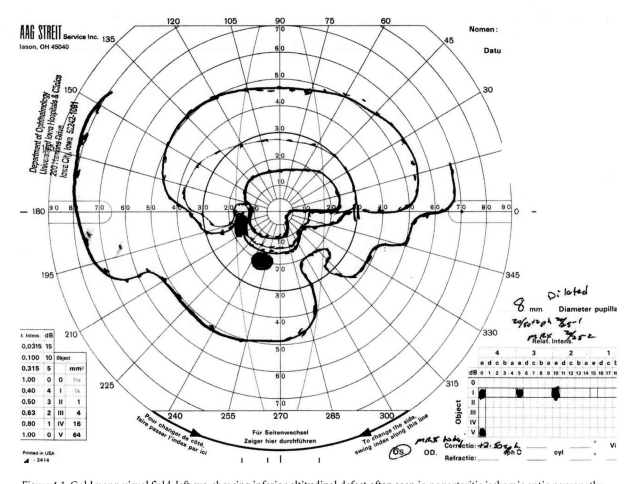

Figure 4.1 Goldmann visual field, left eye, showing inferior altitudinal defect often seen in nonarteritic ischemic optic neuropathy.

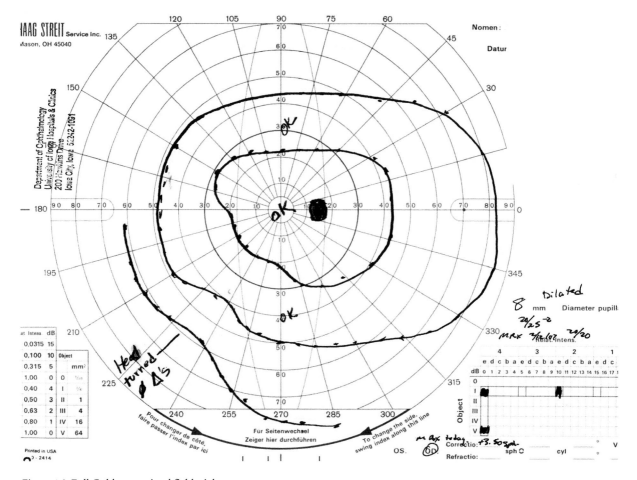

Figure 4.2 Full Goldmann visual field, right eye.

hypotension, diabetes mellitus, cigarette use, hypercholesterolemia, hypertriglyceridemia, elevated fibrinogen, small cup-to-disc ratio, hypercoagulable states, acute blood loss, anemia, elevated intraocular pressure, migraine, sleep apnea, and postcataract surgery.(2–9) Debate exists regarding the exact relationship between NAION and many of these entities.

Systemic hypertension is present in 35–50% of patients with NAION and diabetes mellitus is present in 24–33%.(2, 3, 10) There is general agreement that both these conditions are risk factors for NAION.

Hayreh has suggested that nocturnal hypotension, particularly when associated with other vascular risk factors, may reduce the optic nerve head blood flow below a critical level and thus precipitate NAION.(4) In support of this theory they reported that 75% of their patients noted the visual loss upon awakening. Furthermore, 24-hour blood pressure monitoring showed that patients with NAION have significantly lower nocturnal blood pressures than controls. Use of antihypertensive agents (particularly at night) further reduced the nadir of blood

pressure compounding this potential factor. However, 41% of patients in the ischemic optic neuropathy decompression trial (IONDT) did not report awakening with visual loss and 17% could not remember the onset.(1) These percentages would be compatible with a normal distribution of onset throughout the day. Additionally, Landau and associates did a case-controlled study of 24-hour blood pressure monitoring and found no difference in nighttime diastolic nadir but they did find a lag in the usual rise in blood pressure in the morning.(11)

Sleep apnea syndrome (SAS) is characterized by recurrent partial or complete upper airway obstruction during sleep. Mojon and associates found 12 of 17 patients with NAION to have SAS which was significantly more than their control group.(5) Similarly, Palombi and associates found 24 of 27 patients with NAION to have SAS which represents a 4.9 risk ratio as compared to the general population.(6) Interestingly, Behbehani and colleagues reported 3 patients who developed NAION while being treated with continuous positive airway pressure for SAS.(12)

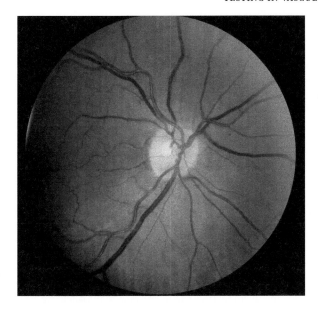

Figure 4.3 Optic nerve, right eye, showing essentially no cup.

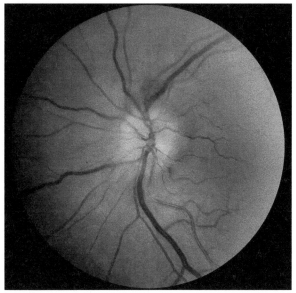

Figure 4.4 Optic nerve, left eye, showing diffuse optic disc edema.

Hypercholesterolemia has been reported to be a risk factor for NAION in several studies.(7, 9, 13) However, a case-control study by Jacobson and associates did not find hypercholesterolemia to be a statistically significant risk factor.(14) Similarly, tobacco use has been reported as a risk factor by some authors (7, 8) but Hayreh (10) and associates reported smoking tobacco was not a risk factor in a series of more than 600 patients with NAION.

No treatment has been found that provides a better visual prognosis than the natural history of the condition.(1) Unfortunately, 15–20% of patients will develop a NAION in the contralateral eye over the following 5 years.(15, 16) There have been no controlled, randomized trials investigating methods to decrease the risk of second eye involvement. Beck and associates conducted a retrospective cohort study on 153 patients treated with aspirin and 278 patients not treated following unilateral NAION.(16) The 2-year probability of developing contralateral NAION was 7% in the treated group and 15% in the untreated group. At 5 years the probability had increased to 15% in the treated group and 20% in the untreated group. This suggests a short-term benefit but this was a retrospective study. Kupersmith and associates also conducted a retrospective study and found aspirin (65–1,300 mg) taken two or more times per week decreased the incidence (17.5% vs. 53.5%) and relative risk ($p = 0.0002$) of second eye AION regardless of the usual risk factors. Salomon and associates retrospectively reviewed 52 patients and also felt there was a benefit of aspirin not only to decrease risk of second eye involvement but also delay onset in eyes ultimately affected.(18) However, in the IONDT, the only prospective study, aspirin use was not found to be a factor in incidence of second eye involvement by NAION.(15)

Thus, there would seem to be little debate that any patient who develops NAION should have their blood pressure, glucose, and cholesterol checked. I do not routinely request 24-hour blood pressure monitoring but if the patient is taking antihypertensive medications, I do counsel them to take the medications in the morning (after obtaining their primary care physician's permission). I ask the patient and spouse about symptoms of sleep apnea and if present I recommend a sleep study. I do not however obtain sleep studies on every patient with NAION. Although tobacco smoking may not be a risk factor for NAION, given its other proven risks the patient should be counseled to quit smoking. I tell the patient the evidence that aspirin use will prevent second eye involvement is poor but these patients are usually in the age range and with other vascular risk factors where aspirin has been shown to have systemic benefits. Thus, I suggest taking one adult strength (325 mg) aspirin per day unless there is some contraindication.

REFERENCES

1. Ischemic optic neuropathy decompression trial research group. Optic nerve decompression surgery for non arteritic anterior ischemic optic neuropathy (NAION) is not effective and may be harmful. JAMA 1995; 273: 625–32.
2. Ischemic optic neuropathy decompression trial research group. Characteristics of patients with non-arteritic anterior ischemic optic neuropathy eligible for the ischemic optic neuropathy decompression trial. Arch Ophthalmol 1996; 114: 1366–74.
3. Hayreh SS, Joos KM, Podhajsky PA, Long CR. Systemic conditions associated with non-arteritic anterior ischemic optic neuropathy. Am J Ophthalmol 1994; 118: 766–80.

4. Hayreh SS. Role of nocturnal hypotension in optic nerve head ischemic disorders. Ophthalmologica 1999; 213: 76–96.

5. Mojon DS, Hedges TR III, Ehrenberg B et al. Association between sleep apnea syndrome and nonarteritic anterior ischemic optic neuropathy. Arch Ophthalmol 2002; 120: 601–5.

6. Palombi K, Renard E, Levy P et al. Non-arteritic anterior ischaemic optic neuropathy is nearly systematically associated with obstructive sleep apnoea. Br J Ophthalmol 2006; 90: 879–82.

7. Talks SJ, Chong NH, Gibson JM, Dodson PM. Fibrinogen, cholesterol and smoking as risk factors for non-arteritic anterior ischaemic optic neuropathy. Eye 1995; 9: 85–8.

8. Chung SM, Gay CA, McCrary JA 3rd. Nonarteritic ischemic optic neuropathy. The impact of tobacco use. Ophthalmology 1994; 101: 779–82.

9. Deramo VA, Sergott RC, Augsburger JJ et al. Ischemic optic neuropathy as the first manifestation of elevated cholesterol levels in young patients. Ophthalmology 2003; 110: 1041–5.

10. Hayreh SS, Jonas JB, Zimmerman MB. Nonarteritic anterior ischemic optic neuropathy and tobacco smoking. Ophthalmology 2007; 114: 804–9.

11. Landau K, Winterkorn JM, Mailloux LU, Vetter W, Napolitano B. 24-hour blood pressure monitoring in patients with anterior ischemic optic neuropathy. Arch Ophthalmol 1996; 114: 570–5.

12. Behbehani R, Mathews MK, Sergott RC, Savino PJ. Nonarteritic anterior ischemic optic neuropathy in patients with sleep apnea while being treated with continuous positive airway pressure. Am J Ophthalmol 2005; 139: 518–21.

13. Giuffre G. Hematological risk factors for anterior ischemic optic neuropathy. Neuroophthalmology 1990; 10: 197–203.

14. Jacobson DM, Vierkant RA, Belongia EA. Nonarteritic anterior ischemic optic neuropathy. A case-control study of potential risk factors. Arch Ophthalmol 1997; 115: 1403–7.

15. Newman NJ, Scherer R, Langenberg P et al. Ischemic Optic Neuropathy Decompression Trial Research Group: the fellow eye in NAION: report from the ischemic optic neuropathy decompression trial follow-up study. Am J Ophthalmol 2002; 134: 317–28.

16. Beck RW, Hayreh SS, Pohajsky PA et al. Aspirin therapy in nonarteritic anterior ischemic optic neuropathy. Am J Ophthalmol 1997; 123: 212–7.

17. Kupersmith MJ, Frohman L, Sanderson M et al. Aspirin reduces the incidence of second eye NAION: a retrospective study. J Neuroophthalmol 1997; 17(4): 250–3.

18. Salomon O, Huna-Baron R, Steinberg DM, Kurtz S, Seligsohn U. Role of aspirin in reducing the frequency of second eye involvement in patients with non-arteritic anterior ischaemic optic neuropathy. Eye 1999; 13: 357–9.

CON: DO NOT TEST FOR BLOOD PRESSURE (NOCTURNAL HYPOTENSION, 24 HOUR BLOOD PRESSURE MEASUREMENTS), SLEEP APNEA, BLOOD SUGAR, CHOLESTEROL, NO SMOKING, ASPIRIN PER DAY

Michael S Lee

The patient presented has a typical clinical story for nonarteritic anterior ischemic optic neuropathy (NAION). Patients with NAION are most commonly in their sixties and do not experience pain. The most common visual field defect is an inferior altitudinal defect like the patient here. The optic nerve is swollen in all cases and nearly all patients have a small cup to disc ratio in the fellow eye. Vision loss can progress for up to a week in many cases of NAION. If the patient demonstrated atypical features, then a workup for other causes may be considered. Suggestive symptoms of arteritic AION include antecedent transient vision loss or diplopia, jaw claudication, headache, malaise, weight loss, anorexia, and scalp tenderness. Suspicious signs include no light perception vision, large cup to disc ratio in the fellow eye, pallid edema, cotton wool spots away from the optic nerve head, and an abnormal temporal artery (tender, pulseless, enlarged). Workup for giant cell arteritis (GCA) should include a Westergren sedimentation rate, C-reactive protein, and a complete blood count. If suspicion is high, then the patient should begin oral corticosteroids 1 mg/kg/day until a temporal artery biopsy. If suspicion is very high, then I will admit a patient for intravenous methylprednisolone 250 mg every 6 hours.

No definitive relationship to carotid disease, heart disease, or stroke exists with NAION. Therefore these patients do not require carotid artery or echocardiographic investigations. Some investigators have suggested that hypercoagulability may cause NAION among young patients. This is not unequivocally established and warrants a word of caution to the clinician—many normal patients demonstrate at least one abnormal laboratory test in the hypercoagulable workup. Additionally, the cost of the workup can run several thousand dollars. I consider a hypercoagulable workup in patients with bilateral simultaneous NAION, recurrent ipsilateral NAION, or a personal/family history of thrombotic events. I do not believe that workup for NAION in patients simply because of young age is high yield. According to the literature sleep apnea appears to occur more frequently among patients with NAION than controls. Cases of fellow eye involvement despite the use of continuous positive airway pressure machines occur and it is not clear that sleep apnea is causative.

There is no evidence that any treatment can improve visual outcome or prevent fellow eye involvement in a patient with NAION. Previous studies have found no advantage to corticosteroids, phenylhydantoin, vasodilators, levodopa, norepinephrine, anticoagulation, or optic nerve sheath fenestration. A couple of large studies have found that the risk of fellow eye involvement does not change with aspirin use or discontinuation of smoking. Brimonidine has shown some neuroprotective properties in animal models of optic nerve damage but remains unproven in humans. A prospective randomized, placebo-controlled, double masked multicentered clinical trial in Europe found that brimonidine did not affect visual outcome.(1) Recent papers have suggested a possible role for radial optic neurotomy, vitrectomy, intravitreal steroids, or bevacizumab; but the number of patients involved is small and the data are not convincing. Finally, the role of nocturnal hypotension in the pathogenesis of NAION is debatable, but I think it is reasonable to ask patients on antihypertensive therapy to take their medications in the morning instead of the evening.

REFERENCE
1. BRAION Study Group. Efficacy and tolerability of 0.2% brimonidine tartrate for the treatment of acute non-arteritic anterior ischemic optic neuropathy (NAION): a 3-month, double-masked, randomized, place-controlled trial. Graefe's Arch Clin Exp Ophthalmol 2006; 244: 551–8.

SUMMARY
Although many hypotheses have been proposed for the etiology of nonarteritic anterior ischemic optic neuropathy (NAION) there remains no proven single cause. Although testing for vasculopathic risk factors seems reasonable (e.g., blood pressure check, evaluation for nocturnal hypotension with 24-hour measurements, sleep study for sleep apnea, blood sugar, cholesterol) there is little evidence that performing these evaluations alters the outcome of the disease. Common sense measures like discontinuation of smoking and consideration for an aspirin per day if there is no contraindication are likewise reasonable but unproven. Unfortunately, there remains no evidence that any evaluation or treatment is effective for NAION.

5   What is the treatment for giant cell arteritis?

An 82-year-old female presents to the local emergency room. She complains of gradual vision loss in the right eye which occurred 2 weeks ago, and more recently has noticed decreasing vision in the left eye as well. She complains of temporal headache, scalp tenderness, and jaw pain after chewing. A CT scan was performed in the ER, which was normal. The ER has already obtained lab tests including a normal CBC, an ESR of 99 mm/hr, and a CRP of 2.7 mg/dl (normal < 0.5). The ophthalmologist is consulted. Her visual acuity is 20/63 in the right eye and 20/50 in the left. There is no relative afferent pupillary defect. She is transported to the eye clinic for additional testing. Abnormal Goldmann visual fields are shown in Figures 5.1 and

5.2. Slit lamp exam is unremarkable. The fundus examination shows disc swelling bilaterally (Figures 5.3 and 5.4). A temporal artery biopsy is scheduled for the next day.

PRO: PATIENTS WITH SUSPECTED GCA AND VISION LOSS SHOULD RECEIVE IV STEROIDS FOLLOWED BY ORAL PREDNISONE AND ANTIPLATELET THERAPY WHILE AWAITING TEMPORAL ARTERY BIOPSY

Timothy J McCulley and Thomas Hwang

The patient outlined above is an ideal example of one that stands to benefit from IV steroids. Although yet to be confirmed by

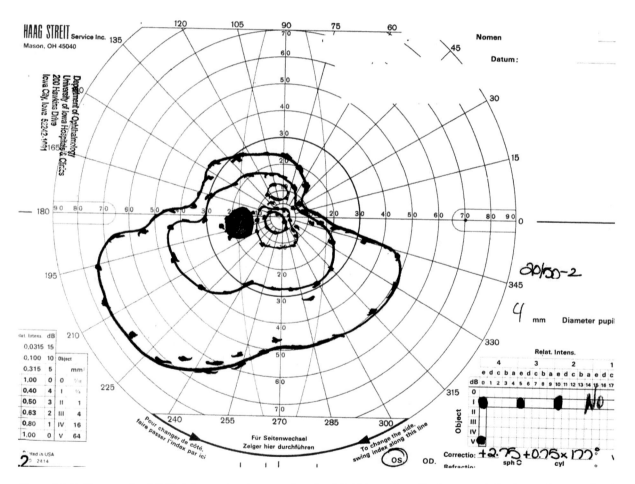

Figure 5.1 Goldmann visual field, left eye, showing superior visual field loss, also affecting central fixation, with generalized constriction as well.

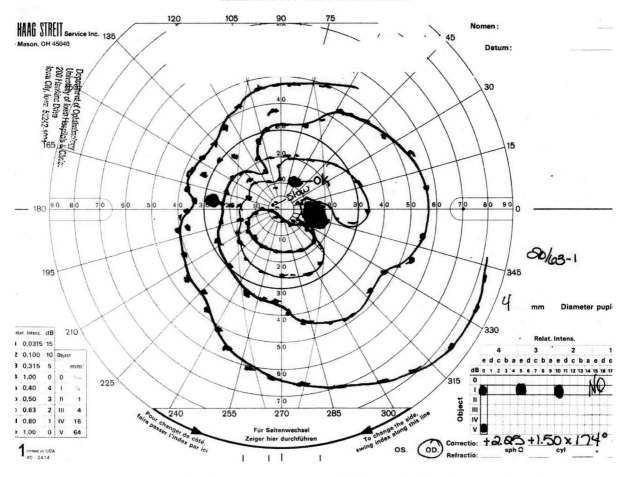

Figure 5.2 Visual field on the right, also showing some superior visual field loss and additional generalized constriction.

biopsy, given the clinical setting and findings the diagnosis of GCA seems highly probable. Substantial visual loss has already occurred in the left eye and there is disk edema and early visual loss in the fellow eye, suggesting impending infarction. IV steroids might result not only in salvage of vision but are most likely to prevent further visual loss in the right eye. Support for this is outlined below.

Theoretically, using IV steroids provides a more rapid and potent antiinflammatory effect than oral steroids alone. It can be argued that by gaining control of the disease more rapidly, IV steroids are more apt to prevent further related complications. Moreover, in the literature, the potential advantages of IV steroids fall into two additional categories, namely improved visual outcome in eyes with AION and shortening of the overall duration of steroid treatment. These will be discussed separately.

In terms of recovery of vision, many anecdotal accounts of visual improvement following high dose IV steroids have been published in the literature over the years. In addition, some retrospective reviews suggest an increased chance for visual improvement with IV steroids. Liu et al. (1993) reviewed 45 biopsy-proven cases of GCA with visual symptoms, 41 of which had visual loss. Twenty received only oral prednisone (40 to 100 mg daily). Twenty-three received IV steroids (250 mg four times daily for 3 to 5 days), but only 13 received it as initial treatment while the remaining 12 received their IV therapy a variable time into their oral prednisone treatment. Although not a statistically significant difference, a higher percentage, 39% (9 out of 23) in the IV steroid group had a measured improvement in Snellen visual acuity compared to 28% (5 out of 18) in the oral steroid group. In support of these results, Chan et al. (2001) later retrospectively reviewed the charts of 100 consecutive patients with biopsy-proven GCA. Patients without visual loss or without adequate medical records or follow-up visits were excluded leaving 73 patients with 43 receiving IV methylprednisolone (dosages ranging from 500 to 1,000 mg for 2 to 5 days) and 30 receiving oral prednisone (dosages ranges from 50 to 100 mg daily).

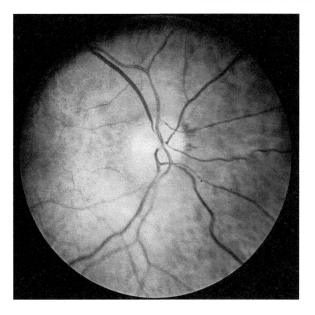

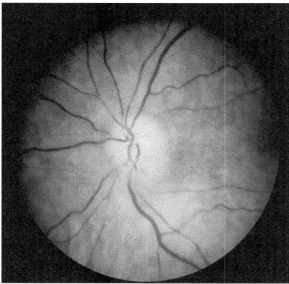

Figure 5.3 Fundus photograph, right eye, showing pallid optic disc edema.

Figure 5.4 Fundus photograph, left eye also showing pallid optic disc edema.

Snellen visual acuity improved in 17 patients (40%) in the IV steroid group *versus* only 4 (13%) in the oral prednisone group ($p = 0.01$). Admittedly, these are anecdotal and/or retrospective and therefore subject to biases inherent in all such reviews. However, they do remain suggestive and are yet to be sufficiently proven an inaccurate reflection of the benefit of IV steroids.

Admittedly, retrospective reviews have been published which failed to show any benefit to IV steroid therapy. Hayreh did not find a statistically significant difference in the number of patients with visual improvement between IV and oral steroids in a retrospective series of 84 consecutive patients with biopsy-proven GCA. However, he used improved central visual field in his definition of improved vision to control for artifactual improvement of Snellen visual acuity from learned use of paracentral vision. Later Hayreh published retrospective data on 144 patients with biopsy-proven temporal arteritis and examined whether IV steroids had an effect on the deterioration of vision in GCA. Both groups had cases of worsening vision during treatment. More patients in the IV steroid treated group experienced further visual loss; however, the proportion was not statistically different that those that did not receive IV steroids. Moreover, it was suggested by Hayreh that the groups were unequal in that patients with more severe disease were more apt to be offered IV steroid therapy. This is the likely explanation for the slight trend towards worsening vision in the IV steroid group. Although these studies failed to confirm benefit from IV steroids, they are insufficiently powered to exclude a clinically relevant effect. Also noteworthy, within the concluding remarks of both manuscripts, it is emphasized that earlier treatment with steroids is preferable, which aligns with the hypothetical benefit of the more rapid-onset IV steroids over oral steroids that just may not have been detected in these retrospective studies.

In terms of a benefit in shorter treatment duration, evidence in the literature shows that initial IV steroids can allow a faster taper of oral steroids with fewer relapses. One study enrolled 27 patients with biopsy-proven GCA in a double-blind, placebo-controlled clinical trial comparing a 3-day course of either IV methylprednisolone (15 mg/kg) or IV saline given once daily while simultaneously starting oral prednisone 40 mg daily. Prednisone was tapered in 2-week intervals from 40 to 10 mg per day over 16 weeks and then tapered by 1 mg per day every 2 weeks. The dose was increased for any clinical or laboratory evidence of relapse and then re-tapered. At 36 weeks, the IV steroid group had 71% of patients (10 out of 14) under 5 mg per day while the placebo group only had 15% (2 out of 13). At 78 weeks, the percentages were 86% and 33% for the IV steroid and placebo groups respectively. The IV steroid group also had fewer relapses during the study period than the placebo group (21 *vs.* 37).

Thus, with regard to IV steroid use in patients with GCA, it is proposed that control of the disease should be achieved as quickly as possible, if for no other reason to prevent further complications. This is applicable to patients with as well as those who have yet to experience visual loss. There is also evidence that reversal of existing visual loss may be more likely in patients treated with IV steroids. Admittedly, evidence in support of this is largely anecdotal and unconfirmed but there has yet to be any data to sufficiently refute the benefits of IV steroids. The main argument against IV steroid use is inconvenience and the unlikely event of a complication; therefore, given both the existing evidence for and lack of a sufficiently powered study to argue against its use, it seems prudent to at least offer IV steroid treatment to patients with GCA.

The second question involves the use of antiplatelet agents in giant cell arteritis as adjuvant therapy. Arguments for the use of antiplatelet therapy are not necessarily specific for the patient presented above but are applicable to all with GCA. The basis of this is the "proven" effectiveness in preventing ischemic events secondary to atherosclerotic disease. The mechanism of this effect is presumably through prevention of thrombus formation in narrowed arteritic vessels that have turbulent flow. Arguably, this would be applicable whether the arterial damage was due to atherosclerosis or GCA. In addition to this theoretical benefit, retrospective studies have suggested utility in using antiplatelet therapies to decrease the rate of GCA-related ischemic events.

In 2004, Nesher et al. published a retrospective chart review of 175 consecutive patients diagnosed with giant cell arteritis. Thirty-six of these were on low-dose aspirin at the time of presentation for cardiac issues. Despite this group having more cerebrovascular risk factors, only 3 patients (8%) had cranial ischemic complications at presentation (cerebrovascular accident [CVA] or vision loss from AION or central retinal artery occlusion) compared to 40 of the 139 patients (29%) not on aspirin ($p = 0.01$). However, not all of the patients in this study had biopsy-proven GCA. Only 34 of the patients (94%) on aspirin and 118 of the nonaspirin patients (85%) had positive temporal artery biopsies. In the remainder, the diagnosis was based on the 1990 American College of Rheumatology (ACR) criteria for GCA. The long-term follow-up of this same cohort was then similarly analyzed. Of the original 175 patients, 9 were lost to follow-up, leaving 166 for this analysis, with a mean follow-up period of 26 months. Seventy-three were treated with prednisone plus aspirin and 93 received only prednisone. Again, despite more cerebrovascular risk factors, the aspirin treated group had statistically better outcomes with only 2 (3%) with cranial ischemic events (1 with vision loss and 1 with CVA) compared to 12 (13%) with events (7 with vision loss and 5 with CVA) in the prednisone only group ($p = 0.02$).

Using a very similar study design, Lee et al. retrospectively identified 143 consecutive patients with GCA based on the 1990 ACR criteria, which included 104 with positive temporal artery biopsies. Sixty-eight were on continuous adjuvant antiplatelet or other anticoagulant therapies during their treatment for GCA. Seventy-five patients did not receive such therapy ($n = 57$) or received it only after having an ischemic event ($n = 18$). Despite having a higher percentage of patients with cerebrovascular risk factors, the patients on adjuvant antiplatelet therapy had statistically fewer ischemic events during the treatment period with 11 of 68 patients (16%) compared to 36 of 75 (48%) in the group without antiplatelet therapy.

Thus, regarding antiplatelet therapy in patients with GCA, there is a theoretical benefit to hindering thrombus formation

within arteries with breakthrough of residual damage due to GCA. And given the supportive published data outlined above, when not contraindicated a strong argument can be made for the use of antiplatelet medications.

We therefore believe that an initial high-dose IV steroid treatment and maintenance with adjuvant antiplatelet therapy are two options available to enhance the standard treatment of GCA with oral steroids. Reasonable theoretical rationales for the use of both can be made and are backed by supportive evidence in form of anecdotal success stories and retrospective analysis that exist in the published literature. Until refuted with an adequately powered prospective study, it seems reasonable to consider offering to patients with GCA related AION, both initial therapy with IV steroids and maintenance with antiplatelet therapy.

BIBLIOGRAPHY

Chan CCK, Paine M, O'Day J. Steroid management in giant cell arteritis. Br J Ophthalmol 2001; 85: 1061–4.

Diamond JP. Treatable blindness in temporal arteritis. Br J Ophthalmol 1991; 75: 432.

Hayreh SS, Zimmerman B. Visual deterioration in giant cell arteritis patients while on high doses of corticosteroid therapy. Ophthalmol 2003; 110: 1204–15.

Hayreh SS, Zimmerman B, Kardon RH. Visual improvement with corticosteroid therapy in giant cell arteritis. Report of a large study and review of literature. Acta Ophthalmol Scand 2002; 80: 353–67.

Jaggi GP, Luthi U, Forrer A, Hasler P, Killer HE. Complete recovery of visual acuity in two patients with giant cell arteritis. Swiss Med Wkly 2007; 137: 265–8.

Lee MS, Smith SD, Galor A, Hoffman GS. Antiplatelet and anticoagulant therapy in patients with giant cell arteritis. Arthritis Rheum 2006; 54(10): 3306–09.

Liu GT, Glaser JS, Schatz NJ, Smith JL. Visual morbidity in giant cell arteritis. Ophthalmol 1994; 101: 1779–85.

Matzkin DC, Slamovits TL, Sachs R, Burde RM. Visual recovery in two patients after intravenous methylprednisolone treatment of central retinal artery occlusion secondary to giant-cell arteritis. Ophthalmol 1992; 99: 68–71.

Mazlumzadeh M, Hunder GG, Easley KA et al. Treatment of giant cell arteritis using induction therapy with high-dose glucocorticoids. Arthritis Rheum 2006; 45(10): 3310–18.

Model DG. Reversal of blindness in temporal arteritis with methylprednisolone. [letter]. Lancet 1978; 1: 340.

Nesher G, Berkun Y, Mates M et al. Low-dose aspirin and prevention of cranial ischemic complications in giant cell arteritis. Arthritis Rheum 2006; 50(4): 1332–37.

Postel EA, Pollock SC. Recovery of vision in a 47-year-old man with fulminant giant cell arteritis. J Clin Neuro-ophthalmol 1993; 13: 262–70.

Rosenfeld SI, Kosmorsky GS, Klingele TG, Burde RM. Treatment of temporal arteritis with ocular involvement. Am J Med 1986; 80: 143–45.

CON: ORAL STEROIDS ARE ADEQUATE TREATMENT FOR GCA

Eric Eggenberger

Giant cell arteritis (GCA) can be a visually devastating disease. Potential sequelae include ischemic optic neuropathy, retinal artery occlusion, and cerebral infarction. Visual loss rarely improves regardless of therapy. Various authors have advocated steroids in differing doses and routes of administration, antiplatelet therapy, anticoagulation, and intraocular pressure lowering agents; however, ideal and evidence-based GCA treatment remains unknown.

Although we typically use high dose steroids in cases of GCA, the route of administration in such cases varies. We have used intravenous (IV) or oral (PO) routes in different cases without evidence-based guidelines and with the knowledge that the IV route does not guarantee visual protection. We reported 4 cases of GCA treated with high dose IV methylprednisolone who subsequently lost vision at least 48 hours into this therapy. Experiences like this emphasize the fact that optimal treatment of GCA remains unknown, and that the IV route does not guarantee the patient freedom from further visual loss. The IV route also presents additional costs and potential complications compared to the oral route. The IV route requires hospital admission or at least skilled administration of the agent through a secure IV line, issues that need to be risk-benefit weighed on an individual basis.

Conversely, the oral route is more convenient and cost-effective. Prednisone and methylprednisolone are both predictably and well absorbed via the oral route. The cost of generic prednisone is a fraction of the expense associated with IV catheter insertion and methylprednisolone administration.

We decide therapeutic route, drug, and dose on an individual patient basis. We will often use the IV route in more urgent cases with recent neurologic, bilateral or severe unilateral visual loss, or progression on oral therapy, realizing this has no evidence based foundation. Until a trial is completed assessing the impact of IV *versus* oral therapy with various dose regiments, the best treatment approach for GCA remains unknown and is decided on a case-by-case basis.

BIBLIOGRAPHY

Cornblath WT, Eggenberger ER. Progressive visual loss from giant cell arteritis despite high-dose intravenous methylprednisolone. Ophthalmol 1997; 104(5): 854–8.

Sayed MH, Al-Habet, Rogers HJ. Effect of food on the absorption and pharmacokinetics of prednisolone from enteric-coated tablets. Eur J Clin Pharmacol 1989; 37: 423–6.

Antal EJ, Wright CE, Gillespie WR et al. Influence of route of administration on the pharmacokinetics of methylprednisolone. J Pharmacokinet Pharmacodyn 1983; 11(6): 561–76.

SUMMARY

There is no "head to head" prospective evidence of superior efficacy for intravenous (IV) versus oral steroids in giant cell arteritis. Although there are risks for IV steroids we believe that it can be offered to selected patients (e.g., monocular, symptoms of transient visual loss, bilateral disease, severe visual loss) as a practice option but there is not sufficient evidence currently to define IV steroids as the "standard of care" for every patient with GCA. Low dose oral aspirin therapy (if there is no contraindication) also has a reasonable biologic rationale but there is insufficient evidence at this time to recommend potentially more dangerous and unproven treatments like anticoagulation with heparin and warfarin. In addition, oral steroids alone have proven sufficient for many of the patients reported in the literature. Thus, we believe that the decision for IV treatment needs to be individualized with the patient.

6 Should I do a bilateral or unilateral temporal artery biopsy in suspected giant cell arteritis?

A 60-year-old man with a past medical history positive for hypertension and hyperlipidemia was referred to the ophthalmology clinic for acute, bilateral visual loss. Two days ago, he noticed multiple black spots in his vision OU, mostly located centrally. He then awoke yesterday morning with almost complete loss of vision OD and progressive loss of his vision OS since then. He has had bilateral frontal headache. He has jaw pain with chewing food, which is more evident at the end of the meal. He has also had a 14 pound unintentional weight loss due to decreased appetite over the past 3 weeks. Visual acuity is only count fingers in each eye. Goldmann visual fields demonstrate only a temporal island of vision OU, slightly larger OS. (Figure 6.1 and 6.2). Dilated fundus examination is shown in Figure 6.3 and 6.4, showing retinal whitening on the right, and optic disc edema on the left. Laboratory testing revealed an erythrocyte sedimentation rate > 140 mm/hr and a C-reactive protein of 22.3 mg/dl (normal < 0.5 mg/dl).

PRO: A BILATERAL TEMPORAL ARTERY BIOPSY SHOULD BE STRONGLY CONSIDERED IN CASES OF SUSPECTED GCA

Michael S Lee

The "gold standard" for the diagnosis of giant cell arteritis (GCA) remains the temporal artery biopsy and in my opinion

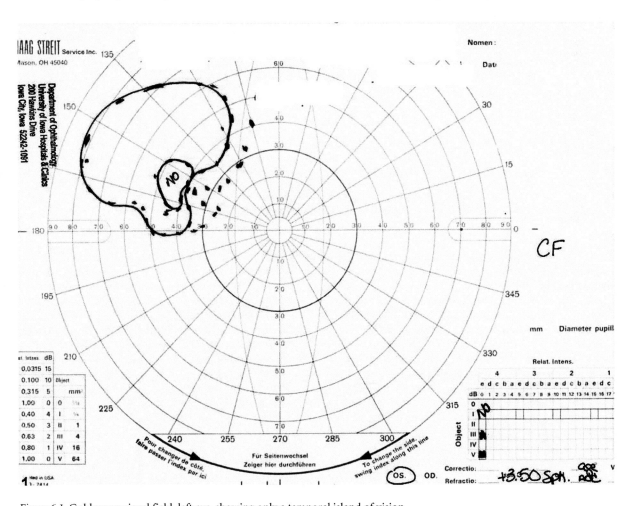

Figure 6.1 Goldmann visual field, left eye, showing only a temporal island of vision.

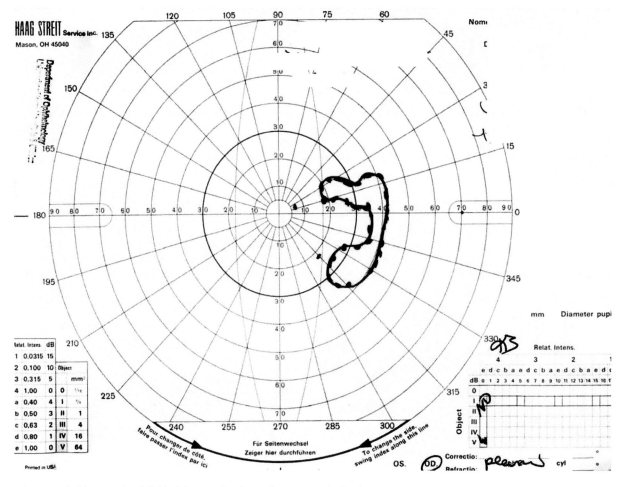

Figure 6.2 Goldmann visual field, right eye, showing only a temporal island remaining.

should be performed in all patients suspected of having GCA. A positive biopsy typically consists of inflammatory mononuclear cells within the vessel walls and disruption of the internal elastic lamina and a positive biopsy result justifies the use of systemic corticosteroid therapy for months or even years. This prevents the premature interruption of therapy when complications of devastating side effects of corticosteroid treatment occur. Since corticosteroids cause the inflammation to disappear, the biopsy should usually be performed within 10–14 days of corticosteroid initiation to avoid a false-negative biopsy.

Giant cell arteritis (GCA) affects medium and large vessels, but it does not cause uniform inflammation of all vessels including both temporal arteries. In some cases, the temporal artery may be unaffected or there may be focal areas of inflammation separated by normal artery, known as skip lesions. A false-negative biopsy result may occur with lack of an adequate tissue sample. Generally a minimum specimen size of 2 cm is recommended to avoid missing the diagnosis. One important

question to ask is "Who is performing the biopsies for you?" I have observed that some general surgeons take only 0.5–0.7 cm of artery, which could easily miss focal inflammation. If your surgeon is providing only small amounts of artery, then a bilateral biopsy would be indicated to yield twice as much tissue and much greater confidence in a negative result.

Three studies have evaluated the role of bilateral temporal artery biopsies vs. unilateral biopsy. The authors determined how often a negative result occurred on one side and a positive result on the opposite side. These studies found that biopsy of the second side would increase the yield between 1 and 5 % over a unilateral biopsy alone. One can reasonably argue that these results represent a minority of cases, but the consequences of both delayed diagnosis of giant cell arteritis leading to bilateral blindness and the misuse of long-term systemic corticosteroids in patients who do not have giant cell arteritis are potentially disastrous. I believe one should have a very low threshold to perform bilateral temporal artery biopsies. It is reasonable to

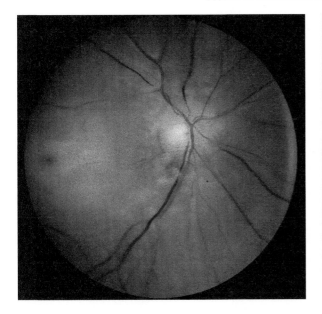

Figure 6.3 Optic nerve photograph, right eye. There is retinal whitening present and nerve fiber layer (NFL) edema. Note the cherry red spot in the macula.

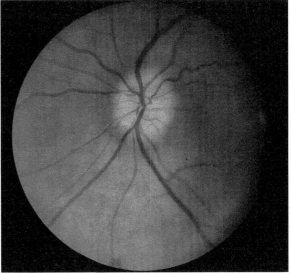

Figure 6.4 Optic nerve photograph, left eye. There is diffuse optic disc edema and pallor.

start with one side, and if clinical suspicion remains high, then the other side should undergo biopsy ideally within 2 weeks after initiating systemic corticosteroids.

CON: A UNILATERAL TEMPORAL ARTERY BIOPSY IS USUALLY ADEQUATE

Wayne T Cornblath

In neuroophthalmology there are very few true emergencies (aneurysmal third nerve palsy, rapidly progressive optic neuropathy from papilledema, thyroid eye disease or pituitary apoplexy, all come to mind). However, the number one neuroophthalmic emergency has to be giant cell arteritis (GCA). GCA has many different manifestations from diplopia to visual loss to scalp lesions, with or without constitutional symptoms. Up to 40% of patients lose vision, and up to three quarters of the 40% lose vision in the both eyes.(1) Yet the treatment of GCA, oral corticosteroids for a year or more, is not without morbidity, particularly given the age range in which GCA occurs. These confounding data points, risk of permanent visual loss *versus* toxicity of treatment, make obtaining a definitive diagnosis of GCA critical.

There are two ways to diagnose GCA, either based on clinical features plus elevated acute phase reactants only or with a temporal artery biopsy (TAB). The clinical criteria frequently cited are those of the American College of Rheumatology. In a review comparing 214 patients with GCA to 593 patients with other forms of vasculitis five criteria were selected:

1) Age greater than or equal to 50 at disease onset
2) New onset of localized headache

3) Temporal artery tenderness or decreased temporal artery pulse
4) Elevated Westergren erythrocyte sedimentation rate
5) Positive TAB

The presence of 3 or more of these criteria lead to a diagnostic sensitivity of 93.5% and the criteria allow a diagnosis of GCA without a temporal artery biopsy TAB.(2) In my practice, the clinical response to prednisone treatment and normalization of blood testing results are also used in decision making. While some have questioned the value of doing a TAB at all and advocate using response to prednisone as diagnostic criteria (3), in general this is a minority view and TAB is viewed as the "gold standard" in diagnosis of GCA.

The question then comes up as to whether to biopsy one side only, both sides sequentially if the first biopsy is negative, biopsy one side and obtain frozen sections and biopsy the second side if the frozen sections on the first side are negative or biopsy both sides simultaneously. Put another way our choices are to biopsy one side only or in some fashion to biopsy both sides.

There are several ways to approach this question. One option would be to say a TAB is a procedure and if we can eliminate a certain percent of procedures with attendant risk and associated costs then this is useful. In that case one could biopsy a single side in all patients and then do a second biopsy in all patients if the first biopsy is negative. If 19–44% of first biopsies are positive (4, 5) then this approach would reduce the number of total biopsies done by 10–22%. For example, 100 patients can undergo 200 simultaneous biopsies or 100 patients can undergo one biopsy and the 56 with negative biopsies undergo a second biopsy for a total of 156 biopsies. Sequential biopsies lead to a 22% reduction (156/200) in the number of biopsies needed. However, this approach, and all approaches with bilateral biopsies, ignores the question of whether a second biopsy actually adds to the diagnostic accuracy. If the second biopsy does not add to diagnostic accuracy then we could reduce the number of procedures by 50%, compared to those who favor bilateral simultaneous biopsies, or a 36% reduction compared to sequential biopsies. So, while all agree on the value of the first biopsy, we must examine the value of the second biopsy.

Before addressing whether the second biopsy is useful we should first review some simple facts about TAB. First, the length of the biopsy is critical. GCA does not affect all portions of the artery equally, producing skip lesions, or segments of normal artery adjacent to abnormal segments. While the percentage of skip lesions is variable, ranging from 8.5% to18% and as high as 28% (6–8) the presence of skip lesions does need to be accounted for. So how long a segment is necessary to eliminate the possibility of a false-negative biopsy? Retrospective reviews have shown that 4, 5 or 10 mm biopsies will avoid the problem of skip lesions.(9–11) While 4 mm might be the minimum length required 15–30 mm is a more desirable range.(12)

Of course, a negative biopsy on one side and positive biopsy on the other side, discussed below, is the "ultimate skip lesion". Specimen shrinkage can occur before excision and after formalin fixation. Su et al measured the artery *in situ* and then after excision and noted an average contraction of 5.7 mm.(13) Danesh-Meyer et al measured an average of 2.4 mm of shrinkage in 54% of specimens after formalin fixation.(14) In addition, crush artifact at the ends of a specimen can also reduce the length available for pathologic review. Given these constraints the 15–30 mm recommendation is reasonable.

Second, adequate processing of the TAB with appropriate sectioning and review by an experienced pathologist is necessary. Wasser reported a case where insurance company insistence on an inexperienced third party laboratory led to a biopsy initially being read as negative that was re-read as positive. In addition, the clinical course clearly supported the diagnosis of GCA.(15) Third, there is an economic cost associated with TAB, with bilateral biopsies obviously costing more than unilateral biopsies. Fourth, on occasion a vein or nerve is biopsied instead of an artery. Ponge reported 9 veins and 3 nerves in 400 biopsies (3%).(16) Boyev noted 4 veins or nerves in 908 biopsies (0.45%).(17) This would lead to a third biopsy in the bilateral biopsy cohort. Fifth, there is a risk of bleeding, infection, facial nerve paralysis, and possibly stroke with TAB. In a discussion of a paper on GCA, C. Miller-Fisher noted a case of stroke during TAB in a patient with ipsilateral carotid occlusion and collateral flow through the external carotid circulation.(18) Sixth, in considering doing sequential biopsies we must consider the effect of additional days of prednisone treatment on the second biopsy result. The common recommendation is to start high dose corticosteroids upon suspicion of GCA and then obtain the

biopsy. If the first biopsy takes 1–3 days to obtain and 3–5 days to process and is negative, a week or more can pass between the first and second biopsy. However, in two reviews with over 600 patients 14–28 days of corticosteroid treatment did not affect the biopsy results.(19, 20) Allowing for 4–8 days from instituting corticosteroids to obtain the first biopsy result and another 3 days to obtain the second biopsy only 7–11 days of corticosteroid therapy have passed, which should not affect the biopsy results. Seventh, some authors recommend frozen section with the first biopsy and if negative simultaneous second biopsy.(21, 17) Unfortunately, a number of biopsies are done in treatment rooms or outpatient facilities where frozen section is not available or practical, so this option has very limited use. Finally, while TAB is considered the gold standard for diagnosing GCA the test is not 100% sensitive, virtually every series has patients with a negative biopsy, or negative bilateral biopsies, who are still felt to have GCA and are treated accordingly. This number can vary from 5% to 44%.(22, 23, 3) In deciding whether to do one biopsy, two sequential biopsies or two simultaneous biopsies these factors must all be considered. For instance, if the first biopsy is only 3 mm in length and is negative, the possibility of skip lesions raises the chance of a false-negative biopsy higher than if the first biopsy was 20 mm in length.

A number of studies have looked at the concordance rate (i.e., agreement of diagnosis between the sides) for bilateral biopsies. The rate of discordance, patients were the first biopsy was negative and the second biopsy was positive is 0–48%. In a large series from the Mayo Clinic there were 234 positive biopsies, of which 201 (86%) were positive with unilateral biopsy. The remaining 33 positive biopsies (14%) were positive on the contralateral, or second, biopsy.(21) A French series of 200 patients with bilateral biopsies had 42 positive biopsies. Twenty were positive bilaterally and 22 were positive unilaterally. This can lead to a discordance rate of 48%, or, assuming half of the 22 unilateral positives would have been found with a unilateral biopsy, a discordance rate of 24%.(16) A large series from Iowa had 363 biopsies with 106 positive biopsies. A subgroup of 76 patients had a second biopsy, 7 of which were positive (9%).(24) A large series from Johns Hopkins had 3 interesting groups. Five hundred and seventy patients had unilateral biopsies, 150 patients had bilateral simultaneous biopsies and 36 patients had sequential bilateral biopsies. In 176 patients the diagnosis on both sides was identical, in 4 patients no artery was obtained on one side. Six patients (3%) had a negative biopsy on one side and a positive biopsy on the other side.(17) A small series of 60 patients with simultaneous and sequential biopsies noted one patient out of 19 in whom a negative first biopsy was followed by a positive second biopsy (5%).(25) Another study of 91 patients with bilateral biopsies had 39 positive biopsies with only one patient having a negative biopsy on one side and positive biopsy on the other side (2.5%). That patient also had small biopsies of 4 and 6 mm, perhaps contributing to the unilateral false negative.(26) As seen a second biopsy can add no additional cases of GCA or up to 48%, with most series ranging from 5–14%.

Most series in the literature of patients with GCA have had only unilateral biopsies but have also treated some patients based on clinical criteria. If a second biopsy was truly required than there should be cases of patients who were not treated and had further complications of untreated GCA or who were diagnosed with GCA after a negative biopsy. Hall et al reported 39 patients with a unilateral negative biopsy followed for an average of 70 months without adverse outcome.(23) Albert et al followed 63 cases with negative biopsy for a minimum of 2 years. Three cases (5%) were felt to have GCA despite negative biopsy. There were no adverse outcomes in 62 of the cases, the 63rd case died of a myocardial infarction 5 months after biopsy. It is not clear if that patient was on prednisone or if the myocardial infarction was related to GCA.(27) Volpe et al reviewed 88 patients who underwent unilateral TAB and were felt to be at low risk for GCA. One patient (1%) had a subsequent second biopsy that was positive and there were no adverse visual or neurologic events in the group.(28) Interestingly from an anecdotal viewpoint most GCA malpractice cases do not involve patients with an initial negative biopsy who then develop complications of GCA but involve patients in whom the diagnosis is not made at the onset of symptoms. These series support the notion that the combination of a single negative biopsy of adequate length plus clinical diagnosis does an excellent job in eliminating or reducing false negatives in the diagnosis of GCA.

Having reviewed the literature we know that a second biopsy will be positive in 0–10% of cases where an adequate length first biopsy is negative and in a similar range of cases the patient will meet the clinical criteria for GCA but will have one or two negative biopsies. We also know that in several series from centers with experienced neuroophthalmologists patients given a diagnosis of "not GCA" do not have high rates of adverse outcomes. Given these facts I would propose the following management for patients with presumed GCA. A unilateral TAB should be done making sure to obtain at least 15 mm and up to 30 mm of artery. If the biopsy is negative and the clinical suspicion is high then a second biopsy can be done. Clinical suspicion, a term much used but little defined, has features we should review. Clinical suspicion should include two areas. First, how compelling are the presenting features of GCA that lead to the biopsy in the first place? Is this a patient with elevated sedimentation rate, fever of unknown origin and no visual symptoms sent for biopsy by the infectious disease service? Or is this a patient with a central retinal artery occlusion in one eye, posterior ischemic optic neuropathy in the other eye 10 days later, new onset severe headache and elevated ESR and CRP? The next feature to consider is the response of symptoms and laboratory findings to corticosteroids. By the time the first biopsy result is available the patient will have been treated with high dose corticosteroids for 3–10 days. Did the clinical symptoms resolve or dramatically improve in 48 hours? Did the elevation in ESR and CRP lessen? By using the combination of an adequate length first biopsy and response to clinical symptoms the decision can be made to proceed with a second biopsy in patients where there

is a high clinical suspicion of GCA. In the Boyev series of 606 patients who did not have initial bilateral biopsies only 36 were felt to need a second biopsy. Similarly, in Hayreh's series of 363 patients only 76 were felt to need a second biopsy. The practice of sequential second biopsies when clinically appropriate will both reduce the number of total biopsies done and not reduce the number of cases of GCA diagnosed.

REFERENCES

1. Birkhead NC, Wagener HP, Shcik RM. Treatment of temporal arteritis with adrenal corticosteroids. Results of fifty five cases in which lesion was proved at biopsy. JAMA 1957; 163: 821–7.
2. Hunder GG, Bloch DA, Michel BA et al. The American College of Rheumatology 1990 criteria for the classification of giant cell arteritis. Arthritis Rheum 1990; 33(8): 1122–8.
3. Allsop CJ, Gallagher PJ. Temporal artery biopsy in giant-cell arteritis. Am J Surg Pathol 1981; 5: 317–23.
4. Stuart RA. Temporal artery biopsy in suspected temporal arteritis: a five year survey. NZ Med J 1989; 102: 431–3.
5. Vilaseca J, Gonzalez A, Cid MC et al. Clinical usefulness of temporal artery biopsy. Ann Rheum Dis 1987; 46: 282–5.
6. Poller DN, van Wyk Q, Jeffrey MJ. The importance of skip lesions in temporal arteritis. J Clin Pathol 2000; 53: 137–9.
7. Albert DM, Ruchman MC, Keltner JL. Skip areas in temporal arteritis. Arch Ophthalmol 1976; 94: 2072–7.
8. Klein RG, Campbell RJ, Hunder GG, Carney JA. Skip lesions in temporal arteritis. Mayo Clin Proc 2007; 82(1): 133.
9. Chambers WA, Bernardino VB. Specimen length in temporal artery biopsies. J Clin Neuro-Ophthalmol 1988; 8(2): 1121–25.
10. Mahr A, Saba M, Kambouchner M et al. Temporal artery biopsy for diagnosing giant cell arteritis: the longer, the better? Ann Rheum Dis 2006; 65: 826–8.
11. Taylor-Gjevre R, Vo M, Shukla D, Resch L. Temporal artery biopsy for giant cell arteritis. J Rheumatol 2005; 32(7): 1279–82.
12. Weyand CM, Goronzy JJ. Giant-Cell Arteritis and Polymyalgia Rheumatica. Ann Interm Med 2003; 139: 505–15.
13. Su GW, Foroozan R, Yen MT. Quantitative analysis of temporal artery contraction after biopsy for evaluation of giant cell arteritis. Can J Ophthalmol 2006; 41: 500–3.
14. Danesh-Meyer HV, Savino PJ, Bilyk JR et al. Shrinkage: Fact or Fiction. Arch Ophthalmol 2001; 119: 1217.
15. Wasser KB. A new skip area in the diagnosis of temporal arteritis. Arthritis Rheum 2005; 53: 147–8.
16. Ponge T, Barrier JH, Grolleau JY et al. The efficacy of selective unilateral temporal artery biopsy versus bilateral biopsies for diagnosis of giant cell arteritis. J Rheumatol 1988; 15: 997–1000.

17. Boyev LR, Miller NR, Green WR. Efficacy of unilateral *versus* bilateral temporal artery biopsies for the diagnosis of giant cell arteritis. Am J Ophthalmol 1999; 128: 211–5.

18. Schlezinger NS, Schatz NJ. Giant Cell Arteritis (Temporal arteritis). Trans Am Neurol 1971; 96: 12–4.

19. Achkar AA, Lie JT, Hunder GG et al. How does previous corticosteroid treatment affect the biopsy findings in giant cell (Temporal) arteritis? Ann Intern Med 1994; 120(12): 987–92.

20. Narvaez J, Bernad B, Roig-Vilaseca D et al. Influence of previous corticosteroid therapy on temporal artery biopsy yield in giant cell arteritis. Semin Arthritis Rheum 2007; 37:13–9.

21. Hall S, Hunder GG. Is Temporal artery biopsy prudent? Mayo Clin Proc 1984; 59: 793–6.

22. Hedges TR, Gieger GL, Albert DM. The clinical value of negative temporal artery biopsy specimens. Arch Ophthalmol 1983; 101: 1251–4.

23. Hall S, Lie, JT, Kurland LT et al. The therapeutic impact of temporal artery biopsy. Lancet 1983; 2(8361): 1217–20.

24. Hayreh SS, Podhajsky PA, Raman R, Zimmerman B. Giant Cell Arteritis: Validity and reliability of various diagnostic criteria. Am J Ophthalmol 1997; 123: 285–96.

25. Pless M, Rizzo III JF, Lamkin JC, Lessell S. Concordance of bilateral temporal artery biopsy in giant cell arteritis. J Neuro-Ophthalmol 2000; 20(3): 216–8.

26. Danesh-Meyer HV, Savino PJ, Eagle Jr. RC et al. Low diagnostic yield with second biopsies in suspected giant cell arteritis. J Neuro-Ophthalmol 2000; 20(3): 213–5.

27. Albert DM, Hedges TR. The significance of negative temporal artery biopsies. Tr Am Ophth Soc 1982; 80: 143–54.

28. Hall JK, Volpe NJ, Galetta SL et al. The role of unilateral temporal artery biopsy. Ophthalmol 2003; 110: 543–8.

SUMMARY

The pre-test likelihood for disease can be used to determine if a bilateral or unilateral temporal artery biopsy (TAB) is likely to yield the diagnosis. In a patient with low clinical suspicion for the diagnosis of giant cell arteritis a bilateral biopsy in our opinion is probably "overkill". Performing a bilateral TAB in every patient is not that technically difficult but does add time to the procedure and is associated with an increased (albeit small) discomfort and surgical risk. The yield of an additional 4% of a bilateral over unilateral TAB has to be considered in the context of the patient. In a patient with high clinical suspicion for the diagnosis, a unilateral TAB followed by a contralateral TAB if the first is negative is reasonable. The only question in this setting is whether the diagnosis from the first TAB can be made at the same sitting (e.g., with frozen section) or at two surgical sessions. The clinician will likely have to balance these issues of time and cost against the pre-test likelihood for disease in specific cases.

7   Should I treat traumatic optic neuropathy?

A 34-year-old male was involved in an altercation during which he was hit on the forehead with a baseball bat. He was believed by his friends to have temporarily lost consciousness, and then was brought to the emergency room due to confusion and swelling over the left eye. The patient was not oriented to place on arrival to the ER, but this problem resolved shortly thereafter. A CT scan showed a small frontal bone fracture overlying the frontal sinus, as well as a small fracture in the roof of the optic canal (Figures 7.1). On examination the visual acuity is reduced to light perception in the left eye and is 20/20 in the right eye. The local ophthalmologist on call is consulted to address this finding. A 3.0 log unit relative afferent pupillary defect OS is identified. Motility exam is normal. Portable slit lamp exam is normal, and direct ophthalmoscopy reveals a normal optic nerve OU (Figures 7.2 and 7.3).

PRO: TREAT TRAUMATIC OPTIC NEUROPATHY WITH HIGH-DOSE STEROID OR POSSIBLY SURGERY

Nicholas Volpe

At this point in time there is no definitive treatment trial to guide decision making in traumatic optic neuropathy (TON). The patient described presumably has fairly isolated posterior indirect traumatic optic neuropathy with severe vision loss. The diagnostic criteria for posterior indirect traumatic optic neuropathy include a nonpenetrating injury with a blow to the face or forehead causing decreased acuity and color vision, a visual field defect, a RAPD in unilateral cases, and normal appearing fundus. In most cases, because the injury is to the facial bones, the globe appears normal with no evidence of traumatic iritis, hyphema, vitreous hemorrhage, or commotio retinae. In fact if there is any evidence of serious eye injury we generally do not recommend treatment of possible TON in conjunction with the eye injury. The differential diagnosis of TON includes preexisting optic neuropathy, retinal compromise secondary to trauma, and functional (nonorganic) vision loss. The incidence of TON is highest in young men (as is any trauma), with bicycle accidents, motor vehicle accidents, and assaults providing the most common setting. Other causes include injuries from falling objects, gunshot wounds, and skateboard related falls. TON can occur after seemingly minor trauma. The incidence is as high as 2–5% after facial trauma.

CT scanning is the diagnostic procedure of choice. Direct, coronal cuts (1.5 mm) (not included in this example) are desirable if the patient can be positioned safely but newer CT coronal reconstructions can also be satisfactory. Coronal CT imaging provides the most detailed views of the optic canal. In the case presented, fractures are identified, and as well there may be bony fragments impinging on the nerve. Occasionally imaging will identify other findings that may be amenable to surgical

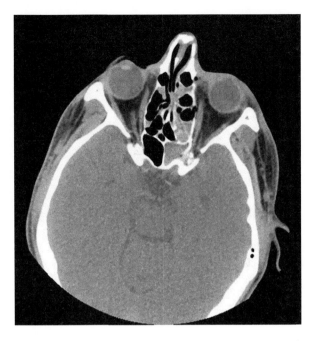

Figure 7.1 CT scan showing small bone fragments near the optic canal on the left.

treatment including an optic nerve sheath or subperiosteal hematoma or hemorrhage in the orbital apex. Identification of a fracture on CT scan is not always possible nor is it a necessary finding to establish a diagnosis of TON. Magnetic resonance imaging (MRI) can be employed to better evaluate soft tissue abnormalities but may not be necessary. CT can also play a role in surgical planning for an optic canal decompression.

As is the case with most patients with this diagnosis, the patient described above is young and has suffered a devastating injury and has many years to live with his lost vision. While there has been no definitive treatment trial, there have been a few studies that have suggested that patients with traumatic optic neuropathy may fair better with treatment with either conventional or mega doses of steroids as well as the possibility of optic canal decompression. The clinician is left making a decision as to whether they are willing to allow for the natural history of this condition to play out or to try and intervene and offer some type of treatment. The recovery may be worse in patients over age 40, with loss of consciousness at the time of injury and with bleeding in the posterior ethmoidal air cells.

Several studies have suggested that the natural history of this condition includes about one-third of patients showing some degree of spontaneous improvement without treatment. There

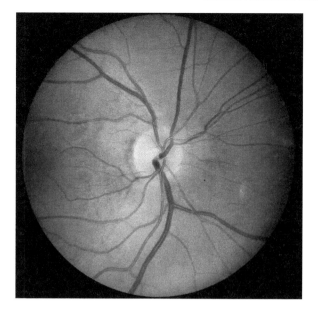

Figure 7.2 normal optic nerve on the right.

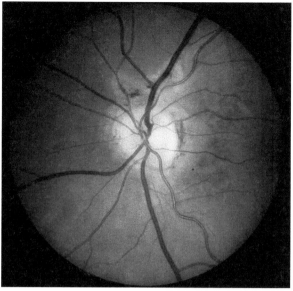

Figure 7.3 Optic atrophy and peripapillary retinal pigment epithelial changes noted in the left eye over time.

are some retrospective studies that would suggest that patients with both conventional, and mega doses of steroids, have an increased likelihood of recovery, perhaps in up to two-thirds of patients. In addition, there are examples within each of these series, including the natural history of patients, standard dose steroid patients and the mega dose steroids patients, in which patients enjoyed dramatic improvements of vision. I believe that even "no light perception vision" is not a contraindication for considering treatment as improvement has been described in these patients.

The presumed mechanism for injury to the optic nerve in indirect traumatic optic neuropathy is thought to occur secondary to mechanical shearing of axons (immediate vision loss) as well as contusion necrosis, perhaps secondary to ischemia and then microvascular compromise. Following a frontal blow, sudden deceleration of the head with continued forward motion of the globe causes shearing forces along the intracanalicular nerve where it has firm attachments to the dura. Additionally, it has been demonstrated in cadaver experiments that the anterior-most portion of the canal, the optic foramen, is the major site of transmitted force from frontal blows. There may be subsequent damage that occurs because of frank swelling of the optic nerve within the optic canal or free radical damage to axons (delayed vision loss). It is likely a combination of apoptotic mechanisms, reperfusion injury, and edema that is responsible for delayed vision loss.

Currently, there is no evidence-based "standard of care" for the treatment of TON. No clear consensus on the efficacy of these treatments has emerged from multiple retrospective or prospective descriptive studies. One of the main sources cited as a rationale for steroid treatment are the National Acute Spinal Cord Injury Studies. These studies investigated steroids for acute brain or spinal cord injury, not specifically TON. The most convincing benefit was seen in the group treated with megadoses of steroids (30 mg/kg followed by a continuous infusion of 5.4 mg/kg/h for 24 or 48 hours) within 8 hours of injury. Admittedly, there is some evidence that steroids may be detrimental. Optic nerve damage has been shown to worsen with steroid administration in animal models. Additionally, results from the CRASH study (Corticosteroid Randomization after Significant Head Injury) suggest that high-dose steroids are associated with increased mortality when given in the context of significant head injury. This large, randomized, placebo-controlled study investigated outcomes following megadose steroid treatment (2 g loading dose followed by 0.4 g/hr over 48 hours) *versus* placebo in 10,008 patients who had suffered head injury. The overall mortality rate 2 weeks following the injury was 21.1% in the steroid group and 17.9% in the placebo group ($p = .0001$). This refutes previous smaller studies that had suggested decreased mortality following steroid treatment for head injury. These results would seem less applicable to our patient who has isolated TON without other significant injury to his brain.

In the end, the clinician is faced with a difficult decision, often based on a limited examination. The clinician can choose to offer these patients steroids for two reasons. The first is that even in conventional doses (IV 250 mg of methylprednisolone four times a day) steroids reduce swelling of the optic nerve within the optic canal and thereby may prevent secondary damage from compression of the optic nerve. Megadoses of steroids, as suggested by the National Acute Spinal Cord Injury Trials, maybe beneficial in patients with traumatic optic neuropathy. In these situations it is presumed that megadoses of steroids help with preventing secondary oxidative damage to the optic nerve. There are however several retrospective studies that have shown steroids to be of no benefit in patients with TON.

If clear evidence of optic neuropathy cannot be obtained because the patient is comatose or uncooperative then no treatment should be offered. However, in a situation where optic nerve damage is identified within 8 hours, is isolated and particularly if it is confirmed to be progressing, treatment should be considered. Here the assumption is that this secondary progression is occurring because of progressive swelling and/or free radical damage to the optic nerve in the tight optic canal and that steroids in both conventional and mega doses should be considered particularly if the rest of the head injury is relatively mild with low risk of significant complications. If the patient continues to progress and/or steroids are ineffective, then optic canal decompression and its ability to prevent secondary damage from swelling within the tight confines of the bony canal, and/or remove offending fractures, (1–3) is another reasonable option to offer patients who are otherwise in a desperate situation with devastating vision loss. It is exactly patients like this, in whom the optic nerve injury is relatively isolated and the risks of head injury and significant complications from the steroids are remote, in whom steroids should be tried as a potential salvage mechanism in an otherwise desperate situation.

REFERENCES

1. Carta A, Ferrigno L, Salvo M., et al. Visual prognosis after indirect traumatic optic neuropathy. J Neurol Neurosurg Psychiatry 2003; 74(2): 246–8.
2. Wang DH, Zheng CQ, Qian J., et al. Endoscopic optic nerve decompression for the treatment of traumatic optic nerve neuropathy. ORL J Otorhinolaryngol Relat Spec 2008; 70(2): 130–3.
3. Wohlrab TM, Maas S, de Carpentier JP. Surgical decompression in traumatic optic neuropathy. Acta Ophthalmol Scand 2002; 80(3): 287–93.

BIBLIOGRAPHY

Levin LA, Beck RW, Joseph MP., et al. The treatment of traumatic optic neuropathy: the International Optic Nerve Trauma Study. Ophthalmology 1999; 106(7): 1268–77.

Lessell S. Indirect optic nerve trauma. Arch Ophthalmol 1989; 107(3): 382–6.

Anderson RL, Panje Gross CE. Optic nerve blindness following blunt forehead trauma. Ophthalmology 1982; 89: 445–55.

Joseph MP, Lessell S, Rizzo J., et al. Extracranial optic nerve decompression for traumatic optic neuropathy. Arch Ophthalmol 1990; 108: 1091–3.

Kline LB, Morawetz RB, Swaid SN. Indirect injury of the optic nerve. Neurosurgery 1984; 14: 756–64.

Seiff SR. High dose corticosteroids for treatment of vision loss due to indirect injury to the optic nerve. Ophthalmic Surg 1990; 21: 389–95.

Spoor TC, Hartel WC, Lensink DB., et al. Treatment of traumatic optic neuropathy with corticosteroids. Am J Ophthalmol 1990; 110(6): 665–9.

Spoor TC, McHenry JG. Management of traumatic optic neuropathy. J Craniomaxillofac Trauma 1996; 2(1): 14–26.

Yu-Wai-Man P, Griffiths PG. Steroids for traumatic optic neuropathy. Cochrane Database Syst Rev 2007; 4: CD006032.

Perry JD. Treatment of traumatic optic neuropathy remains controversial. Arch Otolaryngol Head Neck Surg 2004; 130(8): 1000.

Steinsapir KD, Seiff SR, Goldberg RA. Traumatic optic neuropathy: where do we stand? Ophthal Plast Reconstr Surg 2002; 18(3): 232–4.

Bracken MB. Methylprednisolone and acute spinal cord injury: an update of the randomized evidence. Spine 2001; 26(24 Suppl): S47–54.

Bracken, M.B., Shepard MJ, Collins WF., et al. A randomized, controlled trial of methylprednisolone or naloxone in the treatment of acute spinal-cord injury. Results of the Second National Acute Spinal Cord Injury Study. N Engl J Med 1990; 322(20): 1405–11.

Braughler JM, Hall ED. Current applications of "high dose" steroid therapy for CNS injury. J Neurosurg 1985; 62: 806–10.

Steinsapir KD, Goldberg RA. Traumatic optic neuropathy: a critical update. Compr Ophthalmol Update 2005; 6(1): 11–21.

Coleman WP, Benzel D, Cahill DW., et al. A critical appraisal of the reporting of the National Acute Spinal Cord Injury Studies (II and III) of methylprednisolone in acute spinal cord injury. J Spinal Disord 2000; 13: 185–99.

Dimitiru, C., et al. Methylprednisolone (MP) fails to preserve retinal ganglion cells and visual function following ocular ischemia in rats. Invest Ophthalmol Vis Sci 2008; 49: 5003–7

Steinsapir KD, Goldberg RA, Sinha S., et al. Methylprednisolone exacerbates axonal loss following optic nerve trauma in rats. Restor Neurol Neurosci 2000; 17(4): 157–63.

Edwards P, Arango M, Balica L., et al. Final results of MRC CRASH, a randomised placebo-controlled trial of intravenous corticosteroid in adults with head injury-outcomes at 6 months. Lancet 2005; 365(9475): 1957–9.

Edwards P, Farrell B, Lomas G., et al. The MRC CRASH Trial: study design, baseline data, and outcome in 1000 randomised patients in the pilot phase. Emerg Med J 2002; 19(6): 510–4.

Roberts I. The CRASH trial: the first large-scale randomized controlled trial in head injury. Corticosteroid Randomization After Significant Head injury. Natl Med J India 2002; 15(2): 61–2.

Roberts I, Yates D, Sandercock P., et al. Effect of intravenous corticosteroids on death within 14 days in 10008 adults with clinically significant head injury (MRC CRASH trial): randomised placebo-controlled trial. Lancet 2004; 364(9442): 1321–8.

Levin LA, Baker RS. Management of traumatic optic neuropathy. J Neuroophthalmol 2003; 23(1): 72–5.

Chuenkongkaew W, Chirapapaisan N. A prospective randomized trial of megadose methylprednisolone and high dose dexamethasone for traumatic optic neuropathy. J Med Assoc Thai 2002; 85(5): 597–603.

Entezari M, Rajavi Z, Sedighi N., et al. High-dose intravenous methylprednisolone in recent traumatic optic neuropathy; a randomized double-masked placebo-controlled clinical trial. Graefes Arch Clin Exp Ophthalmol 2007; 245(9): 1267–71.

Steinsapir KD. Treatment of traumatic optic neuropathy with high-dose corticosteroid. J Neuroophthalmol 2006; 26(1): 65–7.

Chen C, Selva D, Floreani S., et al. Endoscopic optic nerve decompression for traumatic optic neuropathy: an alternative. Otolaryngol Head Neck Surg 2006; 135(1): 155–7.

Chen CT, Huang F, Tsay PK., et al. Endoscopically assisted transconjunctival decompression of traumatic optic neuropathy. J Craniofac Surg 2007; 18(1): 19–26.

CON: THERE IS NO PROVEN TREATMENT FOR TRAUMATIC OPTIC NEUROPATHY

Eric Eggenberger

Traumatic optic neuropathy (TON) is a relatively frequent cause of visual loss. Despite this frequency, the natural history is not well defined, and there are no accepted or evidence based guidelines for treatment of TON, and accordingly many therapies have been proposed including medications and surgical decompression.

The untreated prognosis of TON is difficult to succinctly relate, perhaps in part because the mechanism, severity, comorbidities, and applied therapies have varied significantly between case series. In an analysis of 28 reports in the literature, Chou et al. reported improvement in 53% of 176 medically treated patients, 46% of 477 surgically treated patients, and 31% of 81 patients without treatment. The relative numbers of cases in each treatment category alone (81 untreated patients compared to 477 surgically treated cases, which is likely quite different from clinical experience) belies the reporting bias involved in such analyses and emphasizes the scarcity of prospective data. Levin et al. studied treatment effect in 133 nonrandomized patients with TON, focusing on steroid therapy, surgical decompression, and untreated patients. After adjusting for baseline visual acuity, there was no difference between these groups regarding visual improvement, nor was dose or timing of steroids found to be an indicator of visual improvement.

Interest in the use of high dose steroids has been in part fostered by trials in spinal cord trauma (National Acute Spinal Cord Injury Studies [NASCIS] II and III), where megadose regiments have been shown to benefit traumatic cord lesions as evidenced by improved function. Balancing these findings in cord trauma are the results of the Corticosteroid Randomization After Significant Head injury (CRASH) trial in closed head injury (CHI). In contrast to the NASCIS, the CRASH trial demonstrated a disadvantage to the use of high dose steroids in CHI with a higher "all cause" mortality rate among steroid recipients compared to placebo (relative risk 1.18; CI 1.09–1.27). This has fueled speculation that high dose steroids may be disadvantageous in TON, although this is without direct supportive evidence. Given the lack of direct randomized trial evidence of TON therapy, the important, yet unknown, inference issue is whether optic nerve trauma is more akin to spinal cord trauma or cerebral trauma. The optic nerve may bear closer resemblance to the spinal cord than the cerebral hemispheres, but resides in part within the cranial vault, and TON is often accompanied by closed head injury.

Accordingly, although high-dose steroids remain one viable options in select cases of isolated traumatic optic neuropathy, there is potential for worse outcomes following steroid therapy, especially if concomitant closed head injury exists. Optimal therapy for isolated TON awaits a randomized clinical treatment trial.

BIBLIOGRAPHY

Chou PI, Sadun AA, Chen YC et al. Clinical experiences in the management of traumatic optic neuropathy. Neuro-ophthalmology 1996; 18: 325–36.

Bracken MB, Shepard MJ, Holford TR et al. Administration of methylprednisolone for 24 or 48 hours or tirilazad mesylate for 48 hours in the treatment of acute spinal cord injury. Results of the third national acute spinal cord injury randomized controlled trial. J Am Med Assoc 1997; 277: 1597–604.

Levin LA, Beck RW, Joseph MP et al. The treatment of traumatic optic neuropathy: the International Optic Nerve Trauma Study. Ophthalmology 1999; 106(7): 1268–77.

Roberts I, Yates D, Sandercock P et al. Effects of intravenous corticosteroids on death within 14 days in 10008 adults with clinically significant head injury (MRC CRASH trial); randomized placebo-controlled trial. Lancet 2004; 364: 132–8.

SUMMARY

There has not been a large, high statistical power, prospective randomized placebo-controlled clinical trial for corticosteroids in the treatment of traumatic optic neuropathy (TON). Although many treatments have been reported with anecdotal success there are risks for both steroids and surgery. In addition, corticosteroid treatment may produce harm (i.e., increased mortality in the CRASH study) and because the treatment remains unproven there can be no "standard of care" recommendation for any route, dose, duration, or type of steroid treatment in TON. Likewise no specific surgical approach can be considered "standard of care" and the timing and indications remain ill defined for TON. We believe that frank and open discussion of the options, risks, and benefits of unproven treatments should be considered in these patients however and that informed consent should be a well-documented and explicit part of the treatment decision.

8 Should I do a MRI and MR venogram in every patient with pseudotumor cerebri?

A 20-year-old college student presents with headaches and intermittent blurry vision. She has been experiencing generalized headaches for the past 4 months, and pulse synchronous tinnitus for the past 2 months. Her headaches are progressively worse and are now constant over the last week. The pain intensity is 8/10. She also described intermittent visual blurring lasting 30 seconds at a time. She has no nausea, vomiting, or fever. She is in good general health and her only medication is an oral contraceptive. She has gained 15 pounds in the past 6 months and is currently 5'4", weighing 170 pounds. On exam, she has 20/25 visual acuity OD and OS. Her visual fields showed enlargement of the blind spot OU (Figures 8.1 and 8.2) and she has bilateral optic disc edema (Figures 8.3 and 8.4). She went to

the local emergency room last week and had a CT scan of her head which was normal (Figures 8.5).

PRO: MRI AND MRV ARE NECESSARY IN THE WORKUP OF POSSIBLE PSEUDOTUMOR CEREBRI

Nicholas Volpe

Modern neuroimaging and particularly the advent of MRI and MR venography have revolutionized the care of neuroophthalmic patients and have helped us to understand many different neuroophthalmic conditions. In the patient such as the one presented with papilledema and presumed pseudotumor cerebri or idiopathic intracranial hypertension (IIH), it is imperative that the workup be complete in trying to identify specific causes of the elevation of

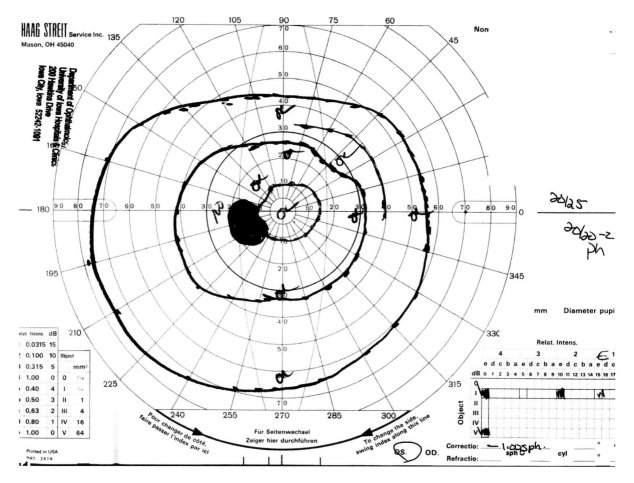

Figure 8.1 Goldmann visual field, left eye, showing an enlarged blind spot.

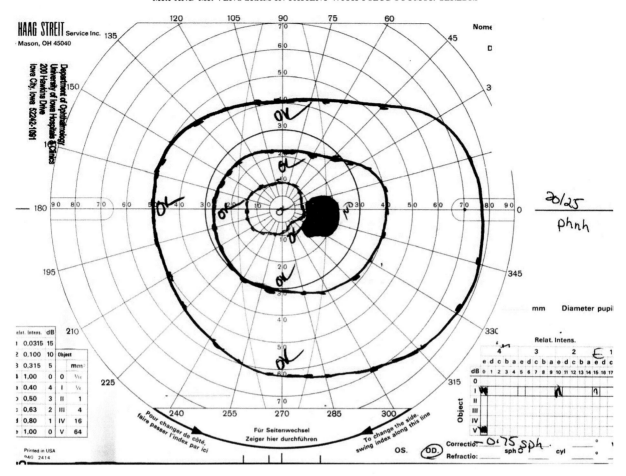

Figure 8.2 Goldmann visual field, right eye, showing an enlarged blind spot.

intracranial pressure. While the vast majority of patients with an otherwise classic presentation, being obese with recent weight gain, will have no specific etiology identified for the elevation of intracranial pressure, there are undoubtedly some patients who will have MRV abnormalities (venous sinus thrombosis) such that this is now included as an important diagnostic criteria for the disease.

Papilledema in sinus thrombosis is indistinguishable from IIH and can occur in patients with both acute and chronic venous sinus thrombosis. One study found 10% of patients who were clinically thought to have IIH turned out to have venous sinus thrombosis. In addition, Biousse et al. clearly defined a subset of patients with venous sinus thrombosis whose presentation was limited to papilledema and very similar to IIH. Obviously it would be even more important to consider this in the differential diagnosis of any patient with papilledema who is a man and/or is a thin woman. These are situations in which IIH is sufficiently atypical that an exhaustive search for lesions such as venous sinus thrombosis and dural arterial venous malformations is mandatory. In addition, there is no reason why a coincidently obese patient could not develop a thrombosis and while the condition is rare and the imaging expensive and technically difficult, the patient is owed

the benefit of the doubt in obtaining this imaging to exclude the possibility of a venous sinus thrombosis. It is imperative that this diagnosis is made in a timely fashion because patients with venous sinus thrombosis can do very poorly. It can be both a life threatening and severely sight threatening condition.

Many patients with venous sinus thromboses progress in a rapid fashion to have life-threatening intracranial thromboses, cortical vein dilation, and venous side strokes. In addition their papilledema can progress to severe ischemic vision loss.

In expert hands, MRI, MRV, and/or CT venography (CTV) are highly specific and sensitive for thrombosis. Some patients with IIH have however been found to have transverse sinus narrowing which is usually readily distinguished from clots and is believed to be a secondary (not causative) epiphenomenon resulting from intracranial pressure elevation. If venous sinus thrombosis is identified in a timely fashion through MRV then there is opportunity to potentially identify an underlying cause for this sinus thrombosis such as dehydration, cancer or a hypercoaguable state, as well as potentially offer treatment, which could include anticoagulation and/or specific interventional radiographic procedures to lyse (e.g., stent, angioplasty) the clots and reopen the system.

41

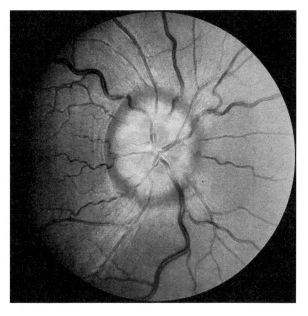

Figure 8.3 Optic disc photo, right eye, showing profound optic disc edema.

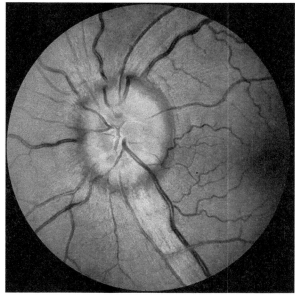

Figure 8.4 Optic disc photo, left eye, also showing marked disc edema.

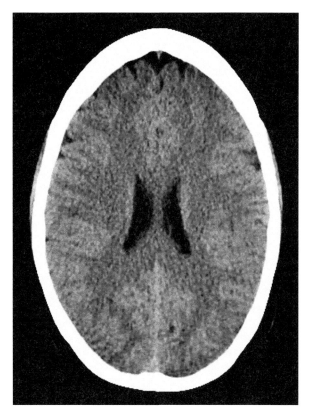

Figure 8.5 Normal brain CT, no evidence of ventricular enlargement or mass.

While MR venography can be technically difficult and there are some conditions that are now likely thought to be largely artifactual, such as narrowing of the transverse sinus, most centers are now able to perform MR venography. Combining the MR venography and the MRI images, which show typical findings when there are clots in the sinuses, the neuroradiographic sensitivity and specificity is high enough for this test that it should be performed in every patient with otherwise unexplained papilledema. The only "down side" of course being the cost of the resource and the potential for misinterpretation, which is unlikely when the study is done carefully and interpreted by experts. Clearly, before offering any type of anticoagulant therapy the MR venography needs to be carefully reviewed and can be combined with both CT angiography and catheter angiography if there is any diagnostic. In the end, venous sinus thromboses are a life threatening cause for papilledema and may cause papilledema that is relentlessly progressing. The clinician is obligated in each patient with papilledema to try and identify this condition and rule it out and/or treat it in an expeditious fashion.

BIBLIOGRAPHY

Friedman DI, Jacobson DM. Diagnostic criteria for idiopathic intracranial hypertension. Neurology 2002; 59(10): 1492–5.

Lin A, Foroozan R, Danesh-Meyer H et al. Occurrence of cerebral venous sinus thrombosis in patients with presumed idiopathic intracranial hypertension. Ophthalmology 2006; 113(12): 2281–4.

Biousse V, Ameri A, Bousser MG. Isolated intracranial hypertension as the only sign of cerebral venous thrombosis. Neurology 1999; 53(7): 1537–42.

Purvin VA, Trobe JD, Kosmorsky G. Neuro-ophthalmic features of cerebral venous obstruction. Arch Neurol 1995; 52(9): 880–5.

de Bruijn SF, de Haan RJ, Stam J. Clinical features and prognostic factors of cerebral venous sinus thrombosis in a prospective series of 59 patients. For The Cerebral Venous Sinus Thrombosis Study Group. J Neurol Neurosurg Psychiatry 2001; 70(1): 105–8.

Agid R, Shelef I, Scott JN, Farb RI. Imaging of the intracranial venous system. Neurologist 2008; 14(1): 12–22.

Khandelwal N, Agarwal A, Kochhar R. Comparison of CT venography with MR venography in cerebral sinovenous thrombosis. AJR Am J Roentgenol 2006; 187(6): 1637–43.

Selim M, Caplan LR. Radiological diagnosis of cerebral venous thrombosis. Front Neurol Neurosci 2008; 23: 96–111.

CON: MRI SCAN WITH CONTRAST IS ADEQUATE IN THE EVALUATION OF POSSIBLE PSEUDOTUMOR CEREBRI AND A VENOGRAM IS USUALLY UNNECESSARY

Fiona Costello

Pseudotumor cerebri (PTC) or IIH is an important clinical diagnosis, with the potential to cause permanent vision loss. Patients are generally overweight, young women who present with headaches, pulse synchronous tinnitus, transient visual obscurations, occasionally, and binocular horizontal diplopia. Patients manifest papilledema but no other localizing neurological signs on examination. This case raises the highly relevant question, regarding what constitutes the evaluation of patients with suspected PTC.

First and foremost, it should be acknowledged that PTC is a diagnosis of exclusion, and tests are therefore done to exclude clinical mimics. The diagnosis of PTC should not be assumed in the case example presented despite the typical demographic profile and clinical characteristics of the patient. Efforts must be exhausted to include all other causes of raised intracranial pressure, including intracranial mass lesions. For this reason, the preliminary study in a patient who presents with features of raised intracranial pressure and bilateral optic disc edema should be a cranial imaging study. Generally speaking, a cranial CT scan can be done in an expedient manner; and is performed, not to make the diagnosis of PTC, but to exclude intracranial hemorrhage, brain tumor or another type of mass lesion. Even in North America but more so in other countries, the rapid access to cranial MRI while preferred is not always possible on demand and therefore a CT scan may be the preliminary imaging study of choice or availability for patients with suspected raised intracranial pressure. In this context, it is not acceptable to wait days to weeks to rule out a potentially life threatening condition. Therefore, unless cranial MRI can be done rapidly, it is not safe to wait.

In the case presented, the patient has already undergone a cranial CT scan, likely for the aforementioned reasons. Important information has been gleaned from this study, including the fact that the patient has no mass lesion, and that the ventricular system appears normal. From a clinical point of view, the patient harbors many of the common risk factors for PTC including female gender, young age, recent weight gain (15 lb), and a body mass index (BMI) that is above her ideal, yet this diagnosis cannot be assumed even at this point. She does not report use of any of the culprit medications that can precipitate raised intracranial pressure including minocycline or vitamin A supplementation. Furthermore, she does not have atypical features including localizing neurological deficits, a low or normal BMI, male gender, prior thrombosis, or systemic symptoms to implicate another potential cause such as meningitis, cerebral venous sinus thrombosis (CVST), or rarely a spinal cord lesion. According to the modified Dandy criteria, she would meet the diagnosis for PTC if she demonstrates an elevated opening pressure (> 25 cm of water) and normal cerebrospinal fluid constituents. For this reason, the lumbar puncture is the next necessary test to perform in this case. The presence of pleocytosis, elevated protein, or positive cultures in the CSF studies would prompt consideration of infectious or inflammatory mechanisms of raised intracranial pressure, and immediately impact clinical management.

The modified Dandy criteria (1) were critically important because they provided clinicians with a step-wise approach to the diagnosis of PTC; however, in the modern imaging era, cranial MRI has become an adjuvant CT imaging (2). MRI provides more detailed images of the brain parenchyma and associated structures than CT; reveals features consistent with raised intracranial pressure (an empty sella, dilated optic nerve sheaths, and flattening of the posterior globe); and can disclose the presence of a Chiari malformation. One potential mimic for PTC, which can be difficult to detect with baseline cranial imaging (enhanced CT or MRI), or clinical criteria alone, is CVST. The prevalence of CSVT is not high, however, and the relatively low risk of this diagnosis probably does not warrant specific imaging of the sinovenous system in all patients. Lin and colleagues performed a retrospective chart review of patients with papilledema from 3 tertiary care neuroophthalmology centers. The occurrence of CVST was 10 (9.4%) of 106 patients with presumed PTC. CVST was diagnosed in 1 of the 10 patients with MRI alone, whereas it was evident in all 10 patients with MR-venography (MRV). The authors concluded that CVST accounts for 9.4% of patients with presumed PTC, and MRI with MRV is recommended to identify this subgroup of patients. While some may argue that all patients with papilledema and suspected PTC should undergo MRV, there are pitfalls to this approach. From a cost-benefit analysis, the study will be unnecessary in approximately 90% of patients. Furthermore, the sensitivity and specificity of MRV in detecting CVST is hindered by false positives, imaging artifacts, and anatomical variants, which can challenge radiological interpretation. Ayanzen and colleagues performed a systematic review of 100 patients with normal MRI studies who underwent MRV. The authors concluded that transverse sinus flow gaps can be observed in as many as 31% of patients with a normal MRI; and that these gaps should not be mistaken for dural sinus thrombosis. Misinterpretation of the MRV could result in mistaken diagnoses of CVST, and cause potential harm to patients secondary to unwarranted anticoagulation or instrumentation. To counter the issues of imaging artifacts with MRV, other modalities of imaging the sinovenous system are often employed including catheter venography or CT-venography (CTV). Khandelwal and colleagues compared CTV and MRV in 50 patients suspected of having CVST. When MRV was used as the gold standard, CTV had a sensitivity and specificity of 75–100% depending on the sinus and vein involved. From their results the authors concluded that CTV is as accurate as MRV.

Despite advances in neuroimaging, PTC remains a clinical diagnosis; and, not, a radiological one. The choice of imaging study therefore needs to be made on a case-by-case basis. In a patient with suspected risk factors for CVST (thrombophilia, critical illness with recent dehydration and weight loss; pregnancy, or an inflammatory condition such as Crohn's disease or ulcerative colitis), sinovenous imaging should be obtained early in the evaluation as the pre-test likelihood for the diagnosis of CVST is higher for these patients than for "all comers" being evaluated for PTC. In the subset of patients at higher risk of CVST, whatever imaging study can be obtained fastest, either MRV or CTV, should be selected to facilitate early diagnosis and management of this important clinical condition. So too, patients with atypical features for the diagnosis of PTC including male gender, advanced age, and low BMI should be thoroughly evaluated for this condition, with MRV, CTV, or catheter venography imaging. Patients with more classic features for the diagnosis of PTC, who maintain well preserved vision and/or respond appropriately to therapy likely do not need imaging of the sinovenous system unless there is a specific indication to do so.

REFERENCES

1. Smith JL. Whence pseudotumor cerebri? J Clin Neuro Ophthalmol 1985; 5:55–6.
2. Friedman DI, Jacobsen DM. Diagnostic criteria for idiopathic intracranial hypertension. Neurology 2002; 59: 1492–95.

BIBLIOGRAPHY

Biousse V, Ameri A, Bousser MG. Isolated intracranial hypertension as the only sign of cerebral venous sinus thrombosis. Neurology 1999; 53: 1537–42.

Lin A, Foroozan R, Danesh-Meyer HV et al. Occurrence of cerebral venous sinus thrombosis in patients with presumed idiopathic intracranial hypertension. Ophthalmology 2006; 113(12): 2281–4.

White JB, Kaufmann TJ, Kallmes DF. Venous sinus thrombosis: a misdiagnosis using MR angiography. Neurocrit Care 2008; 8(2): 290–2.

Alper F, Kantarci M, Dane S et al. Importance of anatomical asymmetries of transverse sinuses: an MR venographic study. Cerebrovasc Dis 2004; 18(3): 236–9.

Ayanzen RH, Bird CR, Keller PJ et al. Cerebral MR venography: normal anatomy and potential diagnostic pitfalls. Am J Neuroradiol 2000; 21: 74–8.

Khandelwal N, Agarwal A, Kochhar R et al. Comparison of CT venography with MR venography in cerebral sinovenous thrombosis. AJR Am J Roentgenol 2006; 187(6): 1637–43.

SUMMARY

There is no doubt that intracranial cerebral venous sinus thrombosis (CVST) can mimic the idiopathic (IIH) version of pseudotumor cerebri (PTC). Most patients however who "fit the profile" (i.e., obese, young females who meet the modified "Dandy" criteria for PTC) do not have venous sinus thrombosis. As tertiary consultants seeing these patients later in their course a neuroimaging study has generally already been performed before our neuroophthalmic evaluation. In our practice if the patient comes with a normal MRI (or less commonly only a CT scan) but no MRV but otherwise meets the modified Dandy criteria, is of the right body habitus and female, and improves with medical therapy and weight loss we often do not make them have a repeat imaging study with MRV. We typically discuss this option with the patient however. In the past there were many flow related artifacts on MR venography without contrast and there was a period of time when this caused a lot of confusion for neuroradiologists and neuroophthalmologists alike. With timed bolus contrast enhanced MRV however most if not all of the prior artifacts have been resolved and so if a patient presents to us with a new presumed diagnosis of PTC we will order a contrast cranial MRI and MRV. On the other hand, we will usually insist on a cranial MRI and MRV for patients who are atypical for PTC including men, younger (i.e., children), and older (i.e., elderly), or thin patients and also for patients who do not follow a typical course of PTC (e.g., severe headache, signs not attributable to increased intracranial pressure alone, or rapid clinical deterioration despite therapy).

Should we perform carotid Doppler and cardiac echo on young patients with transient visual loss?

A 39-year-old previously healthy female presents with episodic decreased vision in the right eye. This has occurred twice so far, and each time the vision appeared to drop out from temporal to nasal. The vision was "almost completely black" each time, and lasted about 3 minutes before resolving gradually over about two more minutes. There were no positive visual phenomena. After the first episode, a mild headache followed which responded to acetaminophen; after the second episode, no headache followed. Both episodes were in the evening. She has a history of migraine in the past with visual aura, but strongly believes that these two episodes are completely different in quality. Her ocular examination reveals visual acuity of 20/20 in each eye, normal visual fields (Figures 9.1 and 9.2), normal pupillary exam, normal slit lamp exam, and normal fundus exam (Figures 9.3 and 9.4). Her OCT shows no evidence of RNFL thinning (Figure 9.5).

PRO: TRANSIENT VISION LOSS IN YOUNG PEOPLE CAN BE THROMBOEMBOLIC SO WORK UP SHOULD BE PERFORMED

Nicholas Volpe

In patients under age 45, the ischemic ocular causes of transient blurring are relatively uncommon. Most of these patients have vasospastic migraine, and almost none of them will go on to develop significant visual or neurologic deficits. That being said, important diagnoses to exclude in young patients with transient

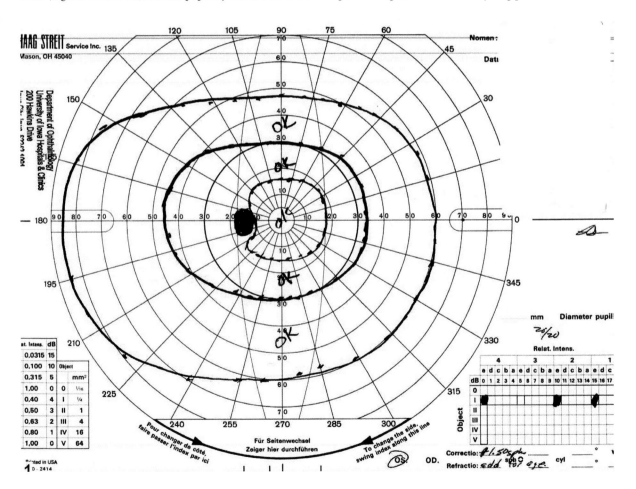

Figure 9.1 Normal visual field on the left.

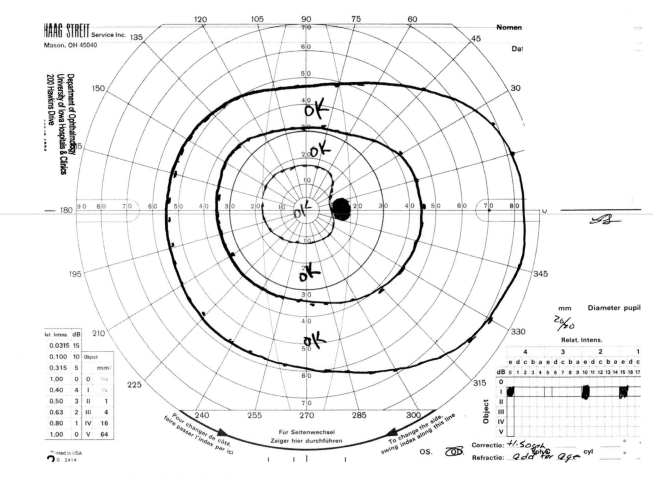

Figure 9.2 Normal visual field on the right.

monocular blindness include: atrial septal defect, cardiac valvular disease, carotid dissection, hypercoagulable states, and connective tissue disorders such as fibromuscular dysplasia. When a clinician is evaluating a patient with a compelling history of transient monocular blindness, which suggests that during the episode the patient lost vision because of poor blood flow to the retinal circulation, there should indeed be an evaluation for treatable causes. While the likelihood of carotid stenosis is low, other conditions such a cardiac valvular disease and patent foramen ovale are diagnostic considerations in this age group, despite the more likely diagnosis of "vasospasm" or "retinal migraine." Admittedly, the group of patients who present with transient visual symptoms is quite heterogeneous and in the majority of cases, thromboembolic causes are not the etiology. Even when it is the etiology, it is rare that a specific treatable entity is found beyond general recommendations to use antiplatelet therapy. However, the clinician would be remiss in my opinion, once a compelling history of vision loss is identified, not to pursue this and once again try and identify causes for this, which potentially maybe amenable to treatment.

The most straightforward situation is the older patient in whom there are classic descriptions of amaurosis fugax or transient monocular blindness. These patients typically develop painless symptoms that last for minutes, and are associated with a sensation of a dark cloud or shade slowing progressing to block the vision in one eye. This type of vision loss, while rarely can be mimicked by a migraine-like condition, is almost always on the basis of some type of thromboembolic phenomenon. Here the stakes are highest for the possibility of significant carotid disease and there is ample evidence to suggest that patients with transient monocular blindness and significant carotid stenosis might indeed benefit from treatment (e.g., carotid endarterectomy). The risk of stroke from transient ischemic vision loss per year had been previously estimated to be 2%, with a 1% risk of permanent visual loss. This compares with the 5% to 8% yearly risk of stroke associated with cerebral transient ischemic attacks (TIAs).

The laboratory evaluation of the patient with suspected ischemic monocular visual loss begins with noninvasive assessment of the

46

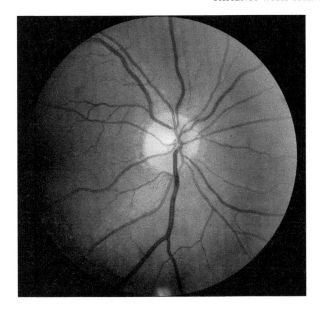

Figure 9.3 Normal right optic nerve.

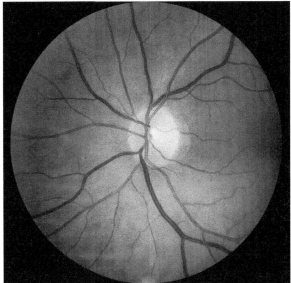

Figure 9.4 Normal left optic nerve.

carotid artery using either ultrasound or magnetic resonance angiography (MRA). Carotid ultrasound and Doppler are effective screening tools for identification and estimation of the degree of internal carotid artery stenosis. This is a "high stakes" clinical decision that should be made with the neuro-ophthalmologist interacting with the endarterectomy surgeon, interventionalist, or the neurologist. There are certainly groups of patients that are even more likely to benefit who have additional risk factors such as older men with smoking history that make the indications for endarterectomy even more compelling. These patients should also be evaluated with echography of the heart as significant cardiac valvular disease with vegetations and/or atrial lesions, such as myxomas or clots and finally, abnormal openings in the cardiac wall such as patent foramen ovale are also in the differential diagnosis of this type of transient vision loss. Finally, in any older patient with transient vision loss, the possibility of temporal arteritis needs to be considered and workup should be directed in such a fashion as to exclude this as a potential cause for transient vision loss particularly in the setting of other constitutional symptoms such as headache, jaw claudication, scalp tenderness, fever, weight loss, or malaise and recurrent symptoms of amaurosis.

In a young patient, such as the one presented, with compelling descriptions of episodes of transient monocular vision loss that lasts for minutes and are characterized by a shade blocking their vision, migraine or vasospasm is the likely etiology, but this remains a diagnosis of exclusion. Vasospastic vision loss may also occur outside of the context of migraine. These patients will have no associated pain or headache, and they may complain of several episodes of monocular visual loss per day. However, retinal vasospasm may occur with increased frequency in patients with connective tissue disorders such

as systemic lupus erythematosus. If a vasospastic cause of the vision loss is suspected, symptoms may be improved with calcium channel blockers.

If necessary, aspirin may also be added to the regimen. Retinal vasospasm and transient monocular visual loss have also been reported in association with exercise and cocaine abuse. While these patients often do not have vasculopathic disease, these patients should be thoroughly evaluated including a transthoracic, and if episodes continue with increased frequency, trans-esophageal echocardiogram (TEE). That is, if the TTE is unrevealing but a cardiac source is still highly suspected, TEE may be useful. TEE has particular advantages over conventional echocardiography in viewing the left atrial appendage, the aorta, and the interatrial septum and in detecting a patient foramen ovale. Carotid evaluation is less critical in this group, although, if there is associated neck pain, then the possibility of a carotid dissection needs to be considered.

There are rare patients who have transient vision loss as an isolated manifestation of hypercoaguable state. This workup should only be extended in patients with previous episodes of spontaneous miscarriages or deep vein thromboses and/or the patient has a compelling history of some type of connective tissue disorder or autoimmune disease. For instance, patients with systemic lupus erythematosus or antiphospholipid antibody syndrome can present with transient vision loss. In a young patient with a confirmed hypercoaguable state and recurrent arterial side ischemic events, anticoagulation should be considered.

While for various reasons it can be argued that transient vision loss has less significant implications then other forms of

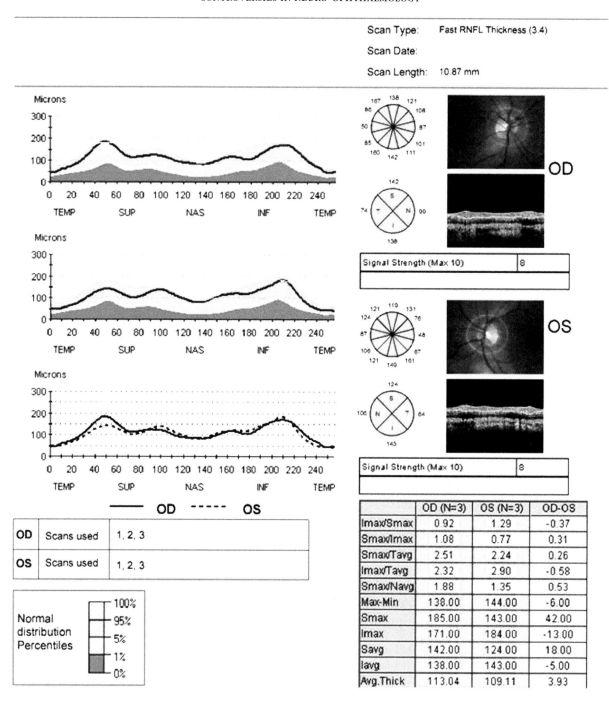

Figure 9.5 Normal OCT of the RNFL.

transient ischemic attacks, such as transient sensory or motor symptoms or transient problems with speech, the clinician must consider a patient with episodic transient vision loss as an individual who is at risk for stroke until proven otherwise. The clinician has a unique opportunity to identify a potential treatable cause for stroke. Although the differential diagnosis of transient vision loss includes many innocent entities, the clinician should not be falsely reassured that their patient is not at risk for stroke. Only when numerous, recurrent, unexplained episodes of visual loss in a young person have been thoroughly investigated, should a diagnosis of migraine or vasospasm be considered.

BIBLIOGRAPHY

Tippin J, Corbett JJ, Kerber RE, Schroeder E, Thompson HS. Amaurosis fugax and ocular infarction in adolescents and young adults. Ann Neurol 1989; 26(1): 69–77.

Slavin ML. Amaurosis fugax in the young. Surv Ophthalmol 1997; 41(6): 481–7.

Poole CJ, Ross Russell RW. Mortality and stroke after amaurosis fugax. J Neurol Neurosurg Psychiatry 1985; 48(9): 902–5.

Wray SH. Visual aspects of extracranial internal carotid artery disease. In: EF Bernstein, ed. Amaurosis Fugax. Springer-Verlag: New York, 1988: 72–80.

Carroll BA. Carotid ultrasound. Neuroimaging Clin N Am 1996; 6(4): 875–97.

Winterkorn JM, Teman AJ. Recurrent attacks of amaurosis fugax treated with calcium channel blocker. Ann Neurol 1991; 30(3): 423–5.

Winterkorn JM, Kupersmith MJ, Wirtschafter JD, Forman S. Treatment of vasospastic amaurosis fugax with calcium channel blockers. N Engl J Med 1993; 329: 396–8.

Jehn A, Frank Dettwiler B, Fleischhauer J, Sturzenegger M, Mojon DS. Exercise-induced vasospastic amaurosis fugax. Arch Ophthalmol 2002; 120(2): 220–2.

Libman RB, Masters SR, de Paola A, Mohr JP. Transient monocular blindness associated with cocaine abuse. Neurology 1993; 43(1): 228–9.

Wisotsky BJ, Engel HM. Transesophageal echocardiography in the diagnosis of branch retinal artery obstruction. Am J Ophthalmol 1993; 115(5): 653–6.

CON: TRANSIENT VISION LOSS IN YOUNG PEOPLE DOES NOT ROUTINELY REQUIRE A CARDIOVASCULAR RISK FACTOR ASSESSMENT

Fiona Costello

There are numerous potential ophthalmic and neuro—ophthalmic causes of transient monocular vision loss and not every patient will require a vascular evaluation. The challenge is to glean the necessary details from the history and physical examination to appropriately direct investigations and identify dire potential causes of transient monocular blindness, while at the same time not embarking on costly, unnecessary investigations.

In this case, the patient is young, healthy, and lacks known vascular risk factors. She does not have a history of known drug abuse, which could predispose her to possible vascular occlusions. She does not report pain or recent trauma to suggest carotid artery dissection. The examination is negative for orbital congestion. Dilated ophthalmoscopy does not demonstrate any areas of vascular occlusion. Furthermore, she has no visual field defects to suggest prior ischemic injury to the afferent visual pathways. The symmetric, and well-preserved retinal nerve fiber layer thickness measured by optical coherence tomography argues against heat—induced conduction block (Uhthoff's phenomenon) as a mechanism of vision loss in this case; which can occur in patients with prior demyelinating insults to the optic nerve. Presumably there is no elevation of her intraocular pressures, and the slit lamp examination is documented as normal.

It is noteworthy that the patient has a prior history of complicated migraines, and experienced a mild headache in association with her first event of vision loss. These historical details raise the possibility that she may be vulnerable to "retinal migraine" or, alternatively, retinal vasospasm. Yet, because there is no specific diagnostic test to confirm migraine as the culprit mechanism of transient monocular vision loss, it remains a diagnosis of exclusion, even in young patients. The definition of what constitutes a "retinal migraine" in clinical practice often deviates from recommendations put forth by the International Headache Society and the term is sometimes loosely applied to all causes of monocular visual disturbance in the young. The danger in this approach is that ischemic mechanisms of visual loss may be missed.

Few would dispute the value of a vascular work—up including carotid Doppler studies, an echocardiogram, and an EKG in the case of an older patient with transient monocular vision loss and established vascular risk factors. Yet, when the patient is young, and lacks vascular risk factors the yield of such investigations is arguably low. In fact, a potential source of emboli is not detected in 50% of patient with retinal arterial occlusive events. Yet, the possibility that this patient harbors a hypercoaguable risk factor or a cardio-embolic source such as valvular heart disease or a patent foramen ovale is not zero. Furthermore, having a history of migraine, which is a ubiquitous condition, does not exclude the possibility that this patient may also have an underlying source of embolism. Because of this she did not experience headache with both episodes of transient visual loss; and given the lack of positive visual phenomena with these events, or a longstanding history of stereotyped events, I would be inclined to err on the side of caution and initiate a vascular work in this case. Patients with altitudinal visual field defects, or lateralized transient monocular vision loss are more likely to have carotid sources of emboli than patients with other patterns of vision loss, and therefore the nature of her clinical presentation would be an for further testing. Because the patient is young, and has no pre-existing risk vascular factors, I would opt for transesophageal echocardiogram (TEE) in lieu of transthoracic echography (TTE) to increase the diagnostic yield. Trans-esophageal echography and TTE were compared in a recent study of 231 consecutive patients with a transient ischemic attack (TIA) or stroke. In this study, TEE proved superior to TTE for identification of cardiac embolic sources in patients with TIA or stroke without pre-existent indication. In patients with normal TTE, a cardiac source of embolism was detected by TEE in approximately 40% of patients, independent of age.

Thus, not all patients with transient monocular vision loss require a complete vascular evaluation, particularly in the absence of vascular risk factors. However, young patients with atypical amaurotic events may require investigations to exclude thrombo-embolism as a potential mechanism for vision loss, because migraine remains a diagnosis of exclusion.

BIBLIOGRAPHY

Kappelle LJ, Donders RC, Algra A. Transient monocular blindness. Clin Exp Hypertens 2006; 28: 259–63.

The International Headache Society International Classification of Headache Disorders, 2nd ed. http://216.25.100.131/upload/CT_Clas/ICHD-IIR1final.doc

Gan KD, Mouradian MS, Weiss E, Lewis JR. Transient monocular vision loss and retinal migraine. CMAJ 2005; 173: 1441–2.

Hill DL, Daroff RB, Ducros A, Newman NJ, Biousse V. Most cases labeled as "retinal migraine" are not migraine. J Neuroophthalmol 2007; 27: 3–8.

Bruno A, Corbett JJ, Biller J, Adams HP Jr, Qualls C. Transient monocular visual loss patterns and associated vascular abnormalities. Stroke 1990; 21: 34–9.

Mouradian M, Wijman CA, Tomasian D et al. Echocardiographic findings of patients with retinal ischemia or embolism. J Neuroimaging 2002; 12: 219–23.

de Bruijn SF, Agema WR, Lammers GJ et al. Transesophageal echocardiography is superior to transthoracic echocardiography in management of patients of any age with transient ischemic attack or stroke. Stroke 2006; 37: 2531–4.

SUMMARY

Most young patients with transient visual loss do not have thromboembolic disease. As the eye exam is typically normal, the key differentiating features in the history should be sought (e.g., monocular altitudinal visual loss "like a curtain", rapid onset in seconds, short duration of minutes) especially in patients with vasculopathic risk factors (e.g., hypertension, diabetes, peripheral vascular disease, older age, prior myocardial infarction or stroke). Younger patients with stereotyped, positive visual phenomenon (i.e., scintillation or fortification scotoma), normal ocular exam, and classic symptoms of migraine aura generally do not require further evaluation. Older and vasculopathic patients with no prior migraine history however with new onset transient visual loss probably deserve consideration for thromboembolic evaluation.

Retinal Migraine

Description:

Repeated attacks of monocular visual disturbance, including scintillations, scotomata or blindness, associated with migraine headache.

Diagnostic criteria:

   A. At least 2 attacks fulfilling criteria B and C.

   B. Reversible monocular positive and/or negative visual phenomena confirmed by examination during an attack by the patient's drawing of a monocular visual field defect during the event.

   C. Headache fulfilling criteria for Migraine without aura, beinning during the visual symptoms or follows within 60 minutes.

   D. Normal ophthalmological examination between attacks.

   E. Not attributed to another disorder.

10   What is the best visual field test for neuroophthalmology?

While working in your office one afternoon you see two patients with visual field loss. The first is an 82-year-old with a longstanding history of primary open angle glaucoma in both eyes. There is advanced glaucomatous cupping, and the intraocular pressure is stable at 16 mm Hg OU on topical timolol and dorzolamide. Previous topical medications have caused allergy and/or were "ineffective". You have followed this patient for several years, but have concern that the Humphrey visual fields (Figure 10.1 and 10.2) may not be providing enough information to allow you to detect progression of disease. A Goldmann visual field is attempted instead at today's visit (Figure 10.3 and 10.4). As you finish seeing the first patient, a second patient is ready to be seen. At the last visit, he had a Goldmann visual field OU to evaluate a recent right occipital lobe infarct (Figure 10.5 and 10.6). At today's visit that test was to be repeated, but a Humphrey visual field was mistakenly performed instead (Figure 10.7 and 10.8). You are surprised to see that the visual field defect is much more apparent at today's visit, even though the patient feels he is doing better.

PRO: GOLDMANN (KINETIC) IS BETTER

Fiona Costello

The choice of visual field modality should be made on a case-by-case basis. The decision to use kinetic perimetry (Goldmann)

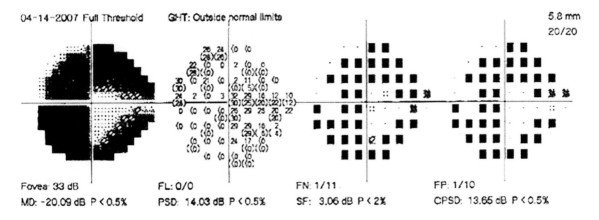

Figure 10.1 Automated (Humphrey 24-2) perimetry, left eye, shows the marked visual field loss.

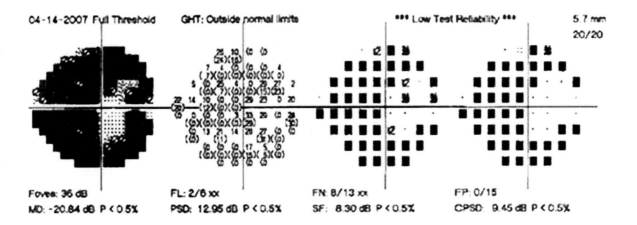

Figure 10.2 Automated (Humphrey 24-2) perimetry, right eye, shows marked visual field loss.

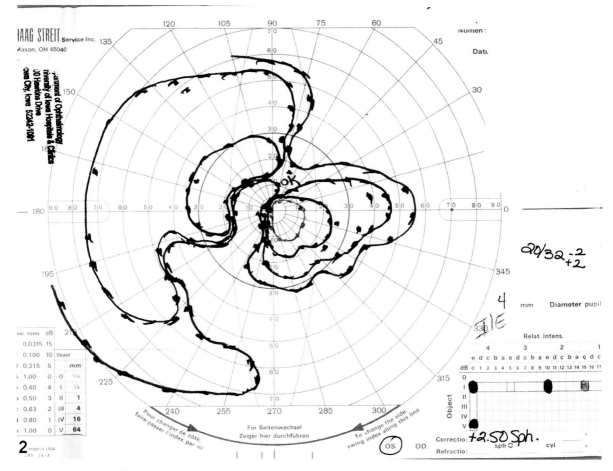

Figure 10.3 Kinetic (Goldmann) perimetry, left eye, shows the preserved peripheral visual field and the extent of the visual field loss better than the automated perimetry in figures 10.1 and 10.2.

versus automated perimetry (Humphrey) depends both on the disease process and the patient being tested. Goldmann perimetry often employs the I4e and I2e isopters in routine ophthalmic practice. Yet, the I1e isopter also provides essential information about the central 10° of visual function, and is essential in the evaluation of neuroophthalmic causes of vision loss. One advantage of Goldmann perimetry is that stimulus presentation is manually controlled, and patients can be instructed and encouraged to do better if they have initial difficulties with testing. This feature can be especially useful in cases where subjects are prone to fatigue or distraction. In addition, Goldmann perimetry does not have a fixed 6° spaced grid, which means that testing can be customized at specific locations to follow regions of interest in the visual field. In complex visual field loss, Goldmann testing (as in the first case of occipital disease) allows characterization of the shape of visual field defects, which helps localize the site, and potential causes, of afferent visual pathway injury. Manual perimetry may be less sensitive than automated perimetry to subtle visual field dysfunction, due to statokinetic dissociation. Furthermore, Goldmann perimetry is more time-consuming and operator dependent than automated perimetry. Therefore, the quality of Goldmann perimetry varies with the skill and expertise of the perimetrist.

Automated perimetry is a clinically practical and less time consuming modality of visual field testing. The stimulus presentation and responses are controlled by a computer, which allows better standardization. Sophisticated statistical programs allow early and sensitive detection of subtle visual field change, without the confounding influence of statokinetic dissociation. Automated perimetry can be difficult to interpret on occasion, because test variability increases with decreasing sensitivity, which means that as the subject's vision worsens it becomes increasingly difficult to detect true visual field change from visual field fluctuation.

In ideal circumstances, I opt to obtain baseline Goldmann perimetry in my initial evaluation of patients with suspected neuroophthalmic causes of vision loss to facilitate topographic

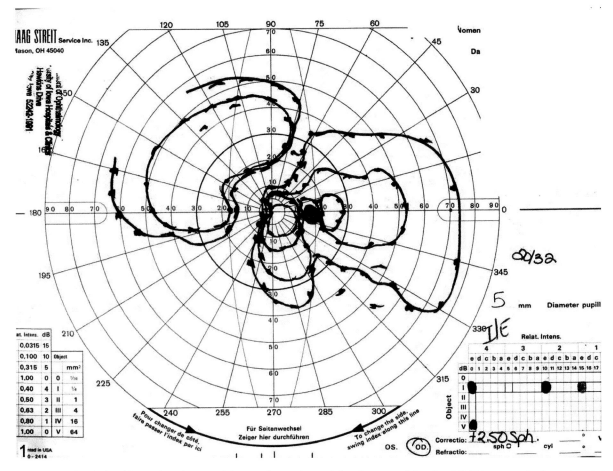

Figure 10.4 Kinetic (Goldmann) perimetry, right eye, shows the peripheral extent of the visual field loss seen in the automated perimetry in figures 10.1 and 10.2.

localization of the visual field defect. Once this is established, I often choose automated (Humphrey) perimetry to follow the effects of disease progression in patients who are able to reliably perform this modality of testing. If automated perimetry results appear highly variable or unreliable, I consult with the perimetrist and the patient to determine why this may be the case. Finally, I review multiple field results to monitor disease progression over time by laying ALL visual field results out sequentially on the floor of my office. In this way, I can determine whether there is a subtle trend toward diminished visual field sensitivity over time, and monitor the test-retest variability for any given patient.

In the first case, an 82-year-old woman with longstanding, primary open—angle glaucoma is followed to look for evidence of disease progression. The Goldmann perimetry results, which include the I1e isopter, provide much more information about the shape and extent of peripheral and central visual field loss due to glaucomatous optic neuropathy in this patient. More specifically, the shape of the visual field defects is quite consistent with glaucoma, and may obviate the need to investigate for

other causes of vision loss. For example, a superimposed left hemianopic defect would be impossible to exclude based on the automated visual fields provided. In the automated perimetry results, the patient demonstrated excessive fixation losses and false-negative responses with right eye testing, which rendered the test unreliable. Excessive false-negative values can be an index of disease severity of in some cases; but in the absence of feedback from the perimetrist, it is impossible to know if the patient fell asleep, maintained a tilted head position, or was unable to comply with testing for some other reason.

In the case of the patient with the right occipital lobe infarct, neither Goldmann nor Humphrey perimetry produced stellar results. The patient has 20/20 vision, but the I1e isopter is not tested with Goldmann perimetry. Therefore, it is unclear whether the patient has left homonymous pericentral scotomas, or a macular splitting homonymous defect. The enlarged blind spots could indicate poor fixation; and statokinetic dissociation could account for the fact that the homonymous field defect was not appreciated with this modality of testing. The comments

53

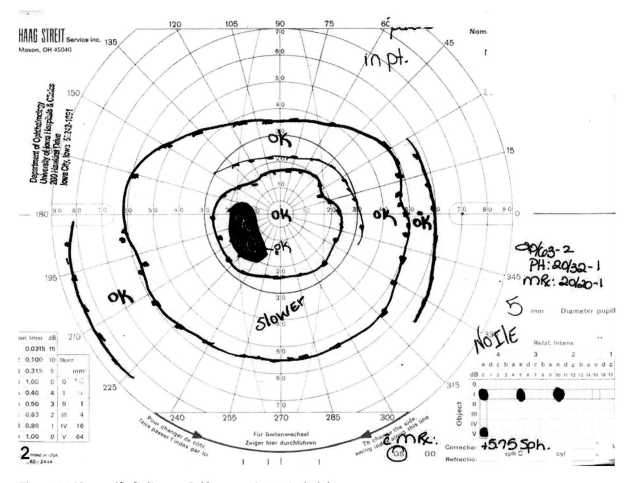

Figure 10.5 Nonspecific findings on Goldmann perimetry in the left eye.

from the perimetrist are sparse, and it is not clear whether the patient was cooperative or inattentive with testing; or whether he was encouraged to give his best performance. The automated perimetry results are also sub-optimal. In the left eye, the foveal threshold is reduced relative to the right, for reasons that are not entirely clear. There are excessive false-negative responses with left eye testing. In the Humphrey visual result for the right eye, there are excessive fixation losses and the blind spot is not properly mapped, which renders this test unreliable. It would be helpful to view the gaze-tracking graphic to determine whether the patient manifested excessive blinking during testing. Furthermore, I would like to know the time required for testing, to determine whether the patient struggled or became fatigued with the process. This case serves to illustrate the point that no test is "best", when the patient or the perimetry have not performed well.

BIBLIOGRAPHY

Levin LA, Arnold AC. Neuro-Ophthalmology: The Practical Guide. New York: Thieme, 2005: 187–97.

CON: AUTOMATED IS BETTER

Wayne T Cornblath

A critical aspect of ophthalmic and neuroophthalmic practice is visual field interpretation. However, before interpretation the type of visual field must be chosen and then adequately performed with the goal of providing reliable information that can be used for either diagnosis (localization) or to follow a process over time. Before the early 80s manual perimetry, usually on a Goldmann perimeter, was the only option for quantifiable, reproducible formal visual field testing. Goldmann perimetry could be time consuming and required a highly trained technician to produce accurate results. With the development of computer automated perimetry another option was available. Studies in the mid 80s were done comparing Goldman perimetry to the Humphrey 30-2 program. The Humphrey program was equal to the Goldmann in showing abnormalities in neurologic patients and in glaucoma and ocular hypertension was more sensitive.(1) Patients preferred the Goldmann and not surprisingly technicians preferred the Humphrey.(2)

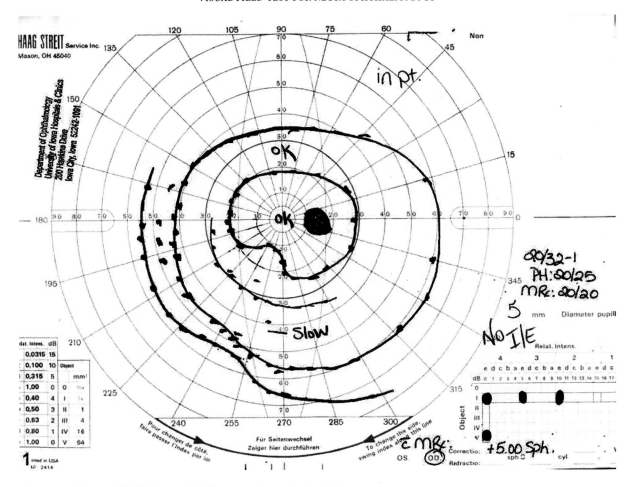

Figure 10.6 Nonspecific findings on Goldmann perimetry in the right eye.

The Humphrey 30-2 program had more fixation difficulties.(1) The Humphrey 30-2 program took an average of 32 minutes for both eyes and the Goldmann took an average of 26 minutes for both.(2) Of course current versions of the Humphrey, particularly the SITA programs, take less time while the Goldmann takes the same amount of time. Studies done a few years later again showed superiority of the Humphrey perimeter compared to the Goldmann for detection of visual field loss in glaucoma (3, 4) and in neuroophthalmic practice.(5)

In a case such as ours with more advanced glaucoma the Humphrey visual field (HVF) appears to be mainly black and of limited usefulness. A Goldmann visual field (GVF) appears to show more information and perhaps should be the field of choice. However, a more discriminating use of the HVF will still provide the information needed to manage this patient and others like him. The first step is to look at all the information on the HVF. When seeing black on the grey scale this indicates a significant difference from age-matched controls, but the patient can still have vision in these areas. We then look at the raw scores and see that in this case the scores are 0 db in a large number of test spots but range from 2 to 20 db in other spots. In comparing future HVFs these raw numbers can be compared point by point for change. The next step before committing this patient to the 40–60 minutes that some glaucoma protocols require (3) is to look at further options with a HVF. The standard 24-2 program is done with a size III isopter, or test target. The isopter can be increased to a size V, a bigger target that is easier to see. A size V target decreases the variability of the test in areas with damage and makes it easier to again see changes over time.(6) In addition, other areas of the visual field can be tested with the Humphrey perimeter, using either additional testing strategies already available in the machine or adding in custom protocols. Pennebaker et al. added a custom temporal periphery program to the standard 30-2 program and then tested glaucoma patients and normal controls. In patients with a relatively normal central field the additional temporal testing added little. However,

55

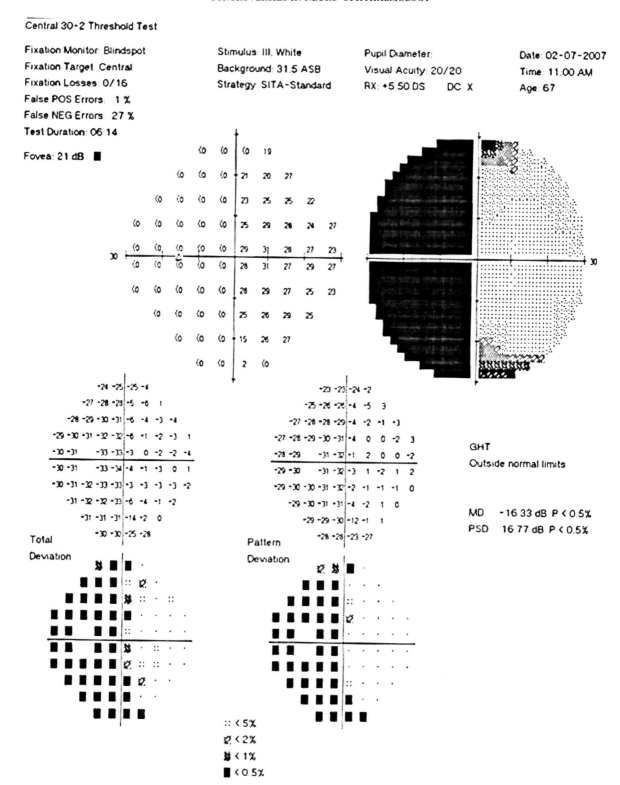

Figure 10.7 Left homonymous visual field loss demonstrated on central 30-2, left eye.

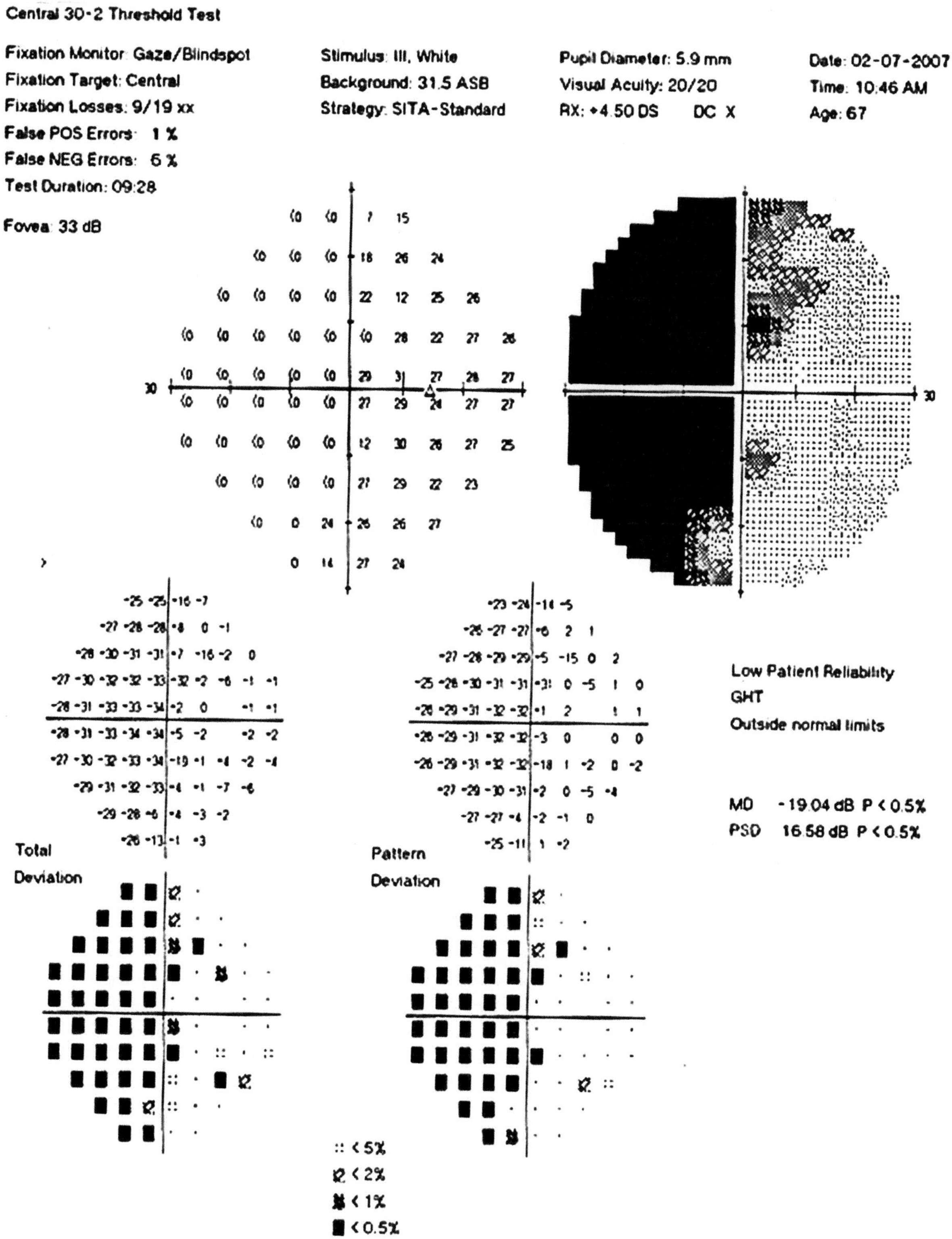

Central 30-2 Threshold Test

Fixation Monitor: Gaze/Blindspot
Fixation Target: Central
Fixation Losses: 9/19 xx
False POS Errors: 1 %
False NEG Errors: 6 %
Test Duration: 09:28

Fovea: 33 dB

Stimulus: III, White
Background: 31.5 ASB
Strategy: SITA-Standard

Pupil Diameter: 5.9 mm
Visual Acuity: 20/20
RX: +4.50 DS   DC X

Date: 02-07-2007
Time: 10:46 AM
Age: 67

Low Patient Reliability
GHT
Outside normal limits

MD   -19.04 dB P < 0.5%
PSD   16.58 dB P < 0.5%

Total Deviation

Pattern Deviation

:: < 5%
⌀ < 2%
✗ < 1%
■ < 0.5%

Figure 10.8 Left homonymous visual field loss demonstrated on central 30-2, right eye.

in patients with a significantly depressed central field, like our case, the addition of the temporal field provided additional area with which to monitor glaucoma progression.

Review of the GVF done shows this area could be followed with a HVF. The final option to monitor this patient would be using an automated perimeter to do a combination of static and kinetic perimetry. Pineles et al. used the Octopus automated perimeter and designed a program to test the central field with static perimetry and test the periphery with kinetic perimetry. This program requires the same level of training to run as standard computer perimetry, i.e., much less training then a Goldmann perimeter. The authors found that their combination program found all the defects that standard kinetic and static testing found along with one additional defect not found with standard testing.(7) Despite a seemingly useless initial 24-2 HVF modifications to the automated perimetry testing in this patient can produce useful information to guide treatment in both less time than the GVF and without the need for a highly trained perimetrist.

The second case illustrates the finding noted in glaucoma and ocular hypertension testing found many years ago: testing with a static threshold method detects more defects than kinetic testing in patients tested with both methods on the same day.(1) The second case also illustrates the importance of choosing the correct HVF protocol. A 30-2 test was done on this patient and by the time the second eye was done the patient had 9/10 fixation losses earning a "low patient reliability" message on the field and appearing to have a new defect in the right homonymous field. The appropriate test for this patient would have been a SITA fast protocol, shown by Szatmary et al. to be equivalent or possibly superior to GVF in patients with similar neuro-ophthalmic visual field defects.(5) If appropriate HVF testing had been done on this patient for his entire course the HVF results would mirror the patient's conclusion of clinical improvement.

Given the prevalence of automated perimeters, the relative dearth of highly trained Goldmann perimetrists and studies showing the comparability of HVF to GVF (with properly chosen programs) the automated perimeter has won the day.

REFERENCES

1. Beck RW, Bergstrom TJ, Lichter PR. A clinical comparison of visual field testing with a new automated perimeter, the Humphrey field analyzer and the Goldmann perimeter. Ophthalmology 1985; 92: 77–82.
2. Trope GE, Britton R. A comparison of Goldmann and Humphrey automated perimetry in patients with glaucoma. Br J Ophthalmol 1987; 71: 489–93.
3. Katz J, Tielsch JM, Quigley HA, Sommer A. Automated perimetry detects visual field loss before manual Goldmann perimetry. Ophthalmology 1995; 102: 21–6.
4. Agarwal HC, Gulati V, Sihota V. Visual field assessment in glaucoma: comparative evaluation of manual kinetic goldmann perimetry and automated static perimetry. Indian J Ophthalmol 2000; 48(4): 301–6.
5. Szatmary G, Biousse V, Newman NJ. Can Swedish interactive thresholding algorithm fast perimetry be used as an alternative to Goldmann perimetry in neuroophthalmologic practice? Arch Ophthalmol 2002; 120: 1162–73.
6. Wall M, Kutzko KE, Chauhan BC. Variability in patients with glaucomatous visual field damage is reduced using size V stimuli. Invest Ophthalmol Vis Sci 1997; 38: 426–35.
7. Pineles SL, Volpe NJ, Miller-Ellis E, et al. Automated combined kinetic and static perimetry: an alternative to standard perimetry in patients with neuroophthalmic disease and glaucoma. Arch Ophthalmol 2006; 124: 363–9.

SUMMARY

The choice of visual field for general ophthalmologists is typically limited by the available perimetry. As automated central computerized perimetry has taken over the market, many ophthalmologists do not have a choice for visual field testing. Although most neuroophthalmic visual field defects will "show up" on the central visual field testing there is still a role in our opinion for manual kinetic (i.e., Goldmann) perimetry. The specific examples of interest to the clinician are the monocular temporal crescent of sparing in occipital lesions that spare the most anterior calcarine cortex and for patients who for a variety of reasons (i.e., elderly patient, child, dementia, poor attention span) do better with direct involvement, perimetrist coaching, and the flexibility of manual perimetry. Automated perimetry for those able to perform a reliable test however provides quantitative reproducible results that can be repeated and compared across multiple tests and multiple testing centers. The reality is that there is no "best visual field test" for every situation and sometimes tests of central 10° (i.e., Amsler grid, Humphrey 10-2 strategy), central 24° or 30° (e.g., Humphrey perimetry), or peripheral field (i.e., Goldmann perimetry) will be superior.

11   Does visual rehabilitation therapy help patients with homonymous hemianopsia?

A 68-year-old male with past history of hypertension presents to the local ophthalmologist on referral from the local neurologist to evaluate visual fields after a presumed stroke. The patient was well until one evening when he became confused and was taken to the emergency room. An MRI of the brain is shown below (Figure 11.1). Based on that finding, neurosurgical consult was obtained, as well as an ophthalmologic consult. On examination, the visual acuity is 20/30 in each eye. Confrontation visual fields indicate a left homonymous hemianopsia; Goldmann visual fields are shown below (Figures 11.2 and 11.3). The remainder of the eye exam is unremarkable. As neurosurgery is planning biopsy of this lesion, the patient asks if anything can be done to help his vision if it does not recover on its own following surgery and planned chemotherapy/radiation therapy.

PRO: VISION REHABILITATION THERAPY CAN BE USEFUL IN CASES OF HOMONYMOUS VISUAL FIELD LOSS

Wayne T Cornblath

In ophthalmic and neuroophthalmic practice homonymous visual field defects are unfortunately very common. Cerebrovascular disease is a common cause of homonymous hemianopia with up to 30% of patients with stroke having a homonymous visual field defect.(1) While some patients will have spontaneous improvement of the visual field defect, the percentage is small and recovery is typically complete by 10–12 weeks.(2) In patients with incomplete recovery disability persists and multiple activities of daily living (ADL) can be impaired. In the state of Michigan, where I practice, 110° of horizontal visual field are required for legal driving vision (tested with Goldmann perimeter). Patients with a complete homonymous hemianopia at best have 90° of horizontal visual field and thus are no longer allowed to drive, a significant disability in many parts of the country where there is not adequate public transportation. Fortunately, there are successful strategies that can be employed to improve functioning and ADLs in patients with visual field defects. Current strategies revolve around either training new scanning techniques or expanding the visual field by lessening the missing area.

When faced with a new visual scene we scan the environment using certain patterns. Patients with a visual field defect of < 6 months duration scan similarly to patients with a normal visual field. After 6 months patients with a visual field defect use different scanning patterns, implying development of a spontaneous compensatory scanning strategy.(3) Observations such as this have lead to efforts to train patients to use more efficient techniques to scan the visual environment, particularly in the area of the visual field defect.

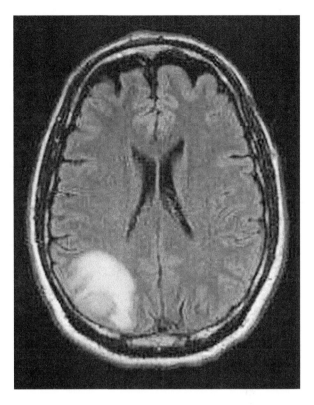

Figure 11.1 MRI showing a hyperintense lesion in the right parieto-occipital lobe on T2 axial fluid attenuation inversion recovery (FLAIR) sequences.

Training new scanning techniques typically involves two techniques: making large saccades into the blind field instead of the usual small saccades and practicing searching techniques on standardized scenes. These techniques are then used in real life situations. When studied, in admittedly small numbers, patients typically show improvement in detection time and reaction time in the hemianopic visual field and show improvement in time required for ADL's. Pambakian et al. showed significant improvement in a group of 29 without a control group by comparing pre and posttraining times.(4) In a group of 21 patients and 23 controls Nelles et al. also showed significant improvement in detection and reaction time and ADL skills.(5) Neither study showed an increase in visual field size after using these techniques. Using a different technique of optokinetic nystagmus (OKN) therapy Spitzyna et al. showed 18% improvement in reading speed in 19 patients treated in a two-armed study.(6)

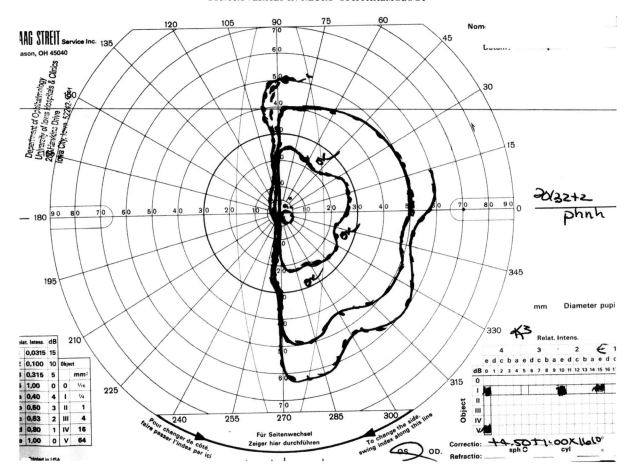

Figure 11.2 Goldmann visual field demonstrating a left homonymous hemianopsia caused by the lesion in Figure 11.1 left eye shown.

In addition, these authors have a free Web-based version of their technique available. For the motivated patient there are techniques to enhance the ability to scan the visual environment and produce meaningful changes in function.

Despite a complete visual field defect, or even removal of the primary visual cortex, patients have been shown to have visual awareness in a blind field, a phenomenon referred to as blindsight.(7, 8) There are several possible explanations for blindsight. Other areas of the brain, such as the superior colliculus or pulvinar could process visual information. Or there could be an element of neuronal plasticity so that other undamaged cortical areas process vision. A recent functional MRI study showed that the visual cortex was activated in sighted patients who were blindfolded and taught Braille. The recruitment of the visual cortex occurred very quickly and then reversed when the blindfolds were removed.(9) These observations have helped form the rationale that lead to the development of training techniques to expand the visual field in the area of scotoma.

In 1998 Sabel et al. published a study showing significant improvement in detection of visual stimuli and 4.9–5.8° expansion in visual angle after patients did computer based visual rehabilitation therapy (VRT).(10) Subsequent studies showed similar results.(11–14) However, concerns were raised that the improvement in visual field related to learning new saccade techniques, as opposed to actual improvement. A subsequent study using scanning laser ophthalmoscope perimetry to control for microsaccades showed no improvement.(15) The original authors then studied VRT improvement while monitoring eye movements and again showed significant visual field enlargement with no effect from eye movements.(16) A study the same year using the Tuebingen Automated perimeter showed no improvement.(17) It is not yet entirely clear whether there is expansion of the visual field with VRT, though a number of studies support this finding. There are also questions about the comparability of standard perimetry which showed improvement and scanning laser

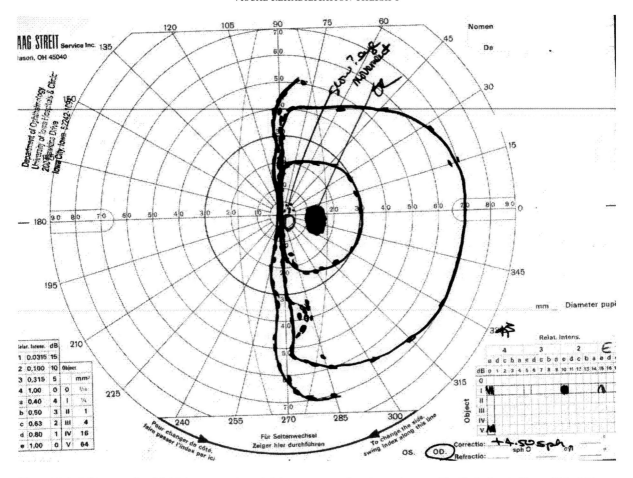

Figure 11.3 Goldmann visual field demonstrating a left homonymous hemianopsia caused by the lesion in Figure 11.1 right eye shown.

ophthalmoscope perimetry which did not show any visual field expansion.(18) Regardless of the findings regarding visual field expansion up to 80% of patients undergoing VRT report improvement in ADL's.(19)

Patients with a homonymous hemianopia and resultant disability can benefit from either training in scanning techniques, VRT or possibly both.

REFERENCES

1. Pambakian AL, Kennard C. Can visual function be restored in patients with homonymous hemianopia? Br J Ophthalmol 1997; 81: 324–8.
2. Gray CS, French JM, Bates D et al. Recovery of visual fields in acute stroke: homonymous hemianopia associated with adverse prognosis. Age Ageing 1989; 18: 419–21.
3. Pambakian AL, Wooding DS, Patel N et al. Scanning the visual world: a study of patients with homonymous hemianopia. Neurol Neurosurg Psychiatry 2000; 69(6): 751–9.
4. Pambakian ALM, Mannan SK, Hodgson TL, Kennard C. Saccadic visual search training: a treatment for patients with homonymous hemianopia. J Neurol Neurosurg Psychiatry 2004; 75: 1443–8.
5. Nelles G, Esser J, Eckstein A et al. Compensatory visual field training for patients with hemianopia after stroke. Neurosci Lett 2001; 306: 189–19.
6. Spitzyna GA, Wise RJS, McDonald SA et al. Optokinetic therapy improves text reading in patients with hemianopic alexia: a controlled trial. Neurol 2007; 68: 1922–30.
7. Stoerig P, Cowey A. Blindsight in man and monkey. Brain 1997; 120: 535–59.
8. Weiskrantz L, Warrington EK, Sanders MD, Marshall J. Visual capacity in the hemianopic field following a restricted occipital ablation. Brain 1974; 97(4): 709–28.
9. Merabet LB, Hamilton R, Schlaug G et al. Rapid and reversible recruitment of early visual cortex for touch. PLoS ONE 2008; 3(8): e3046.

10. Kasten E, Wust S, Benrens-Baumann W, Sabel BA. Computer-based training for the treatment of partial blindness. Nat Med 1998; 4: 1083–7.

11. Kerkhoff G. Restorative and compensatory therapy approaches in cerebral—a review. Restorative Neurol Neurosci 1999; 15: 255–71.

12. Kasten E, Poggel DA, Miller-Oehring E et al. Restoration of vision II: residual functions and training-induced visual field enlargement in brain-damaged patients. Restorative Neurol Neurosci 1999; 15: 273–87.

13. Sabel BA, Kenkel S, Kasten E. Vision restoration therapy (VRT) efficacy as assessed by comparative perimetric analysis and subjective questionnaires. Restorative Neurol Neurosci 2004; 22: 399–420.

14. Romano JG, Schulz P, Kenkel S, Todd DP. Visual field changes after a rehabilitation intervention: vision restoration therapy. J Neurol Sci 2008; 273: 70–4.

15. Reinhard J, Schreiber A, Schiefer U et al. Does visual restitution training change absolute homonymous visual field defects? A fundus controlled study. Br J Ophthalmol 2005; 89(1): 30–5.

16. Kasten E, Bunzenthal U, Sabel BA. Visual field recovery after vision restoration therapy (VRT) is independent of eye movements: an eye tracker study. Behav Brain Res 2006; 175: 18–26.

17. Schreiber A, Vonthein R, Reinhard S. Effect of visual training on absolute homonymous scotomas. Neurol 2006; 67: 143–5.

18. Glisson C. Capturing the benefit of vision restoration therapy. Curr opin ophthalmol 2006; 17: 504–8.

19. Mueller I, Poggel DA, Kenkel S, Kasten D, Sabel B. Vision restoration therapy after brain damage: subjective improvements of activities of daily life and their relationship to visual field enlargements. Vis Impair Res 2003; 5: 157–78.

CON: VISION REHABILITATION THERAPY IS NOT HELPFUL IN CASES OF HOMONYMOUS VISUAL FIELD LOSS

Eric Eggenberger

Patients with homonymous hemianopia often have a difficult time adapting to loss of vision. In many ways, this is a more challenging deficit than monocular loss of vision. This is especially true concerning the ability to drive following such lesions; homonymous hemianopia prohibits driving, while monocular loss of vision with an intact fellow eye allows driving privileges. Because of these factors, several investigators have advocated various rehabilitation strategies, however, results vary depending upon the specific difficulty and the technique applied.

Perhaps the most visible of the visual rehabilitation strategies is the computer-based Vision Restoration Therapy (VRT, NovaVision AG). This therapy claims to improve visual field deficits via cortical plasticity. Although there is no known medical risk

of the therapy, there are several potential disadvantages of this therapy. Treatment is typically quite expensive, reportedly in the $6,000–7,000 range for full course, and this cost is born by the patients (not insurance covered). The visual field improvement outcome is measured by NovaVision's proprietary perimetry despite the well-established computerized perimetry strategies that are available in virtually every ophthalmology office; this unfamiliar outcome tool renders the results less understandable and generalizable. Studies using Tuebingen Automated Perimetry (TAP) and scanning laser ophthalmoscope (SLO) perimetry have not convincingly demonstrated evidence of expanded perimetry post-treatment. Additional criticism of VRT surrounds lack of control for saccades into the blind hemifield, eye movements that could be learned during therapy and improved apparent visual function in the lost hemifield. In contrast, there are other methodologies that may assist patients with certain visually-based deficits at little or no cost. One technique has been applied to hemianopic alexia, in which a right homonymous hemianopia impairs reading. Spitzyna and colleagues demonstrated improved reading speed following a computer-based scrolling print therapy (available free on line at http://www.readright.ucl.ac.uk/).

Thus, it is my opinion that visual rehabilitation is in its infancy. The clinician should be vigilant that rehabilitation strategies employed by their patients first do no harm; injury in this sense may take medical or financial forms. Furthermore, such strategies require repeat independent confirmation applying standard outcomes before general acceptance by the neuroophthalmology community.

BIBLIOGRAPHY

Spitzyna GA, Wise RJS, McDonald SA et al. Optokinetic therapy improves text reading in patients with hemianopic alexia: a controlled trial. Neurology 2007; 68: 1922–30.

Reinhard J, Schreiber A, Schiefer U et al. Does visual restitution training change absolute homonymous visual field defects? A fundus-controlled study. Br J Ophthalmol 2005; 89: 30–5.

Schreiber A, Vonthein J, Reinhard S et al. Effect of visual restitution training on absolute homonymous scotomas. Neurology 2006; 67: 143–5.

Glisson CC, Galetta SL. Visual rehabilitation. Now you see it; now you don't. Neurology 2007; 68: 1881–2.

SUMMARY

Patients with stable partial or complete homonymous hemianopsias often have significant functional deficits (e.g., reading, avoiding running into objects in their nonseeing field, driving). Visual rehabilitation may have a role in selected patients by improving saccadic search strategies into the blind field and perhaps in a few patients by cortical plasticity mechanisms. Although the evidence is encouraging for "vision restoration" therapy the "jury is still out" on the efficacy and cost effectiveness as well as the mechanism for the subjective and objective improvements that have been reported to date.

12   Should a patient with a pupil involved third nerve palsy have a catheter angiogram if the MRA or CTA are negative?

A 52-year-old man presents to the ophthalmologist with a past medical history of diabetes type II, asthma, hyperlipidemia, coronary disease with prior stent placement, and chronic low back pain. He also has a 50 pack/year history of tobacco use. Three days ago he began noticing that his right upper eyelid was droopy and that he was experiencing double vision. The ptosis progressed and is now almost complete. He is complaining of global headache, without nausea or vomiting. On examination, there is mydriasis OD with partial ptosis of the right upper lid. Ocular motility revealed on the right side, a -4 adduction deficit and a -3 elevation and depression deficit (Figure 12.1). He was seen at an outside hospital 2 days ago and an MRI/MRA were reportedly "normal".

PRO: A PATIENT WITH A PUPIL-INVOLVED THIRD NERVE PALSY SHOULD HAVE AN ANGIOGRAM IF THE MRA OR CTA ARE NEGATIVE

Timothy J McCulley and Soraya Rofagha

Choosing the appropriate evaluation for a patient with an isolated third nerve palsy can be one of the most challenging decisions faced by an ophthalmologist, often leaving the clinician with a sense of doubt and unease. Historically the debate about imaging arose because one had to decide whether or not to obtain an invasive angiogram but now there are minimally invasive or noninvasive imaging such as magnetic resonance imaging (MRI) and now MR angiography (MRA) or CT

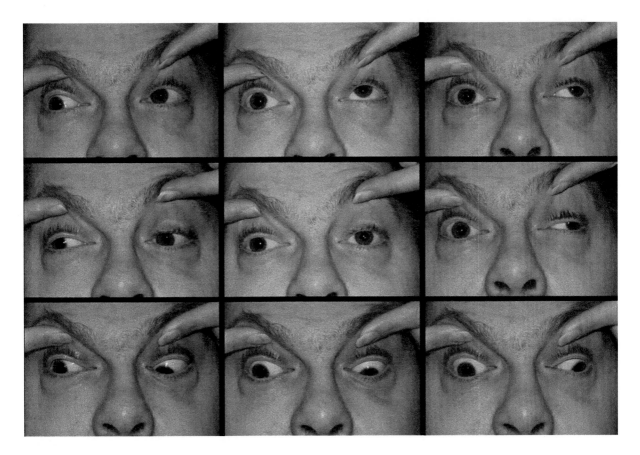

Figure 12.1 Motility photographs showing an adduction, elevation, and depression deficit on the right, with an exotropia and right hypotropia consistent with a third-nerve palsy. Careful examination also reveals a larger pupil on the right.

angiography. Many neurologists (and neuroophthalmologists) have advocated that an initial imaging study is not required in every neurologically *isolated, pupil spared, complete* third cranial nerve palsy in a vasculopathic patient but the controversy remains. Historical features such as an HIV positive patient or a history of a neoplastic or lymphoproliferative disorder would not be considered isolated cases however.

Conventional CT might be useful for the evaluation of subarachnoid hemorrhage but cranial MRI is superior for the evaluation of nonaneurysmal causes of an isolated third nerve palsies. The real dilemma has been when angiography should be obtained and what type of "angiography". Digital subtraction angiography (DSA) commonly referred to as "conventional catheter angiography", caries a small but inarguable risk for complication, up to 5% risk in older populations. It is because of this risk that catheter angiography has traditionally been reserved for those with a higher probability of harboring an aneurysm signified by a number of potentially differentiating features (e.g., partial nerve palsy, pupil involvement, younger age and a lack of vascular risk factors). Given the number of variables and consequently the inconsistency among patients, no consensus or specific "formula" regarding who does and who does not require angiography has been agreed upon. Patients are approached individually.

As stated above, in the past, the diagnosis of an aneurysm was depended largely on invasive conventional catheter cerebral angiography. Advances in MR and CT angiography (MRA and CTA), however have changed the diagnostic evaluation of patients with isolated third nerve palsies. With the availability of MRA and CTA either of which can be obtained with minimal risk, the threshold for recommending catheter angiography, for most clinicians, is now lower. We agree that it is reasonable to use MRA/CTA as a screening tool in the evaluation of patients suspected of having an aneurysm. As a consequence of this practice, we are occasionally faced with patients in whom our suspicion for an aneurysm is high, but have a "*normal*" (or more commonly technically inadequate or otherwise suboptimal) MRA or CTA. The question then becomes whether or not conventional angiography is indicated, in light of a "negative" MRA/CTA. Because of the "high stakes" nature of the decision, we argue that until it is established that CTA and/or MRA are as or more sensitive and specific than conventional angiography, DSA will continue to play a role in the evaluation of patients with third nerve palsies.

In 2000, guidelines published by *The Stroke Council* recommend intra-arterial catheter angiography as the "gold standard" for detection of an intracranial aneurysm. Regarding CTA, with its resolution being limited to 2–3 mm, they advised its use in follow-up of a known aneurysm. Similarly, MRA with a resolution limited to 3–5 mm was recommended only as screening tool. However, standard, single-slice CTA (the previously dominant technology) has been largely replaced with helical or multidetector scanner technology, which has markedly improved spatial resolution. Ultimately, CTA may prove to be equally sensitive/specific as conventional angiography. But a proper comparative study has yet to demonstrate this.

MAGNETIC RESONANCE ANGIOGRAPHY

Several MRA techniques may be employed: a gadolinium MRA, 2-dimentional time of flight (2D TOF), or 3-dimentional time of flight (3D TOF). Intracranial vasculature is most often assessed with 3D TOF. This technique is reliant of blood flow within the vessel and depending on a vessel's particular flow characteristics, might underestimate its caliber. This also holds true for aneurysms, which may be missed entirely. There is sufficient data available to establish DSA as being superior to MRA. Numerous studies have shown that MRA identifies only a fraction of intracranial aneurysms, which were detected with DSA. For example, in 2001, White and colleagues prospectively evaluated 142 patients with MRA, CTA, and DSA. MRA detected 86% of aneurysms viewed with DSA that were > 5 mm in diameter and 35% of aneurysms < 5 mm in size. Using the previously reported sizes of aneurysms producing third nerve palsies, the relationship of aneurysm size and risk of rupture and known MRA sensitivity, Jacobson and Trobe estimated that MRA will miss 1.5% of aneurysms responsible for third nerve palsies that are likely to rupture.

COMPUTED TOMOGRAPHY ANGIOGRAPHY

CTA is more sensitive than MRA but still may not be as good as catheter angiography in all locations. One drawback of CTA is that bone artifact may obscure visualization of an aneurysm and conventional angiography does not share this shortcoming. In the prospective study conducted by White and colleagues, CTA was found to be more sensitive than MRA but still not as sensitive as DSA. Although CTA detected 94% of aneurysms viewed with DSA that were greater than 5 mm in diameter, only 57% of aneurysms less than 5 mm in size were identified. In a blinded prospective study comparing DSA and CTA in the setting of acute subarachnoid hemorrhage, sensitivity and specificity of spiral CTA was estimated to be 86% and 90%, respectively. The six falsely negative CTA in this study were attributed to small aneurysms size (<4 mm). In an experimental model, Piotin and colleagues also found CTA to be more accurate than MRA but still not as sensitive as DSA. In a meta-analysis of twenty-one references, calculations based on data for 1,251 patients who underwent CTA resulted in a sensitivity of 93%. Numerous other studies have yielded similar results. Borrowing the formula utilized by Jacobson and Trobe in the assessment of MRA, assuming 100% sensitive in detecting aneurysms greater than 5mm, and generously estimating sensitivity to be 75% for aneurysms less than 5 mm, we can estimate that roughly 0.5% of aneurysms responsible for third nerve palsy likely to rupture if left untreated would be missed.

The patient described above has a near complete isolated third nerve palsy with minimal pupil involvement. He also has multiple vascular risk factors and moreover, known atherosclerotic disease. As outlined above, an aneurysm has not been entirely excluded with the normal MRA. However, in this case our suspicion of an aneurysm is relatively low and it would be difficult to argue for conventional angiography. As discussed above a quality CTA would increase sensitivity over an MRA and might be a reasonable consideration at this time. Should a CTA fail to identify an aneurysm, close observation would probably be appropriate. Conventional angiography may still have a role. But it should be reserved in our opinion for cases where the palsy progresses or fails to resolve as expected of a microvascular cranial nerve palsy.

In conclusion, catheter angiography in our opinion remains the gold standard for evaluation of unruptured aneurysms. MRI/MRA is not sufficient to exclude third nerve palsy producing aneurysm. Modern CTA offers improved sensitivity relative to MRA and may eventually surpass DSA. To date this has not been substantiated and unless demonstrated adequately in a comparative study, CTA should also not be considered a substitute for DSA. The use of MRA or CTA as a screening tool seems reasonable. The risk of missing an aneurysm that will progress to rupture is near and possibly less than the risk of serious injury (stroke, myocardial infarction or death) with DSA. However, the consequences of a ruptured aneurysm exceed that of a complicated angiogram, and these numbers cannot be directly compared. In patients more highly suspect of having an aneurysm, a negative CTA or MRA should still be followed with DSA.

BIBLIOGRAPHY

Bederson JB, Awad IA, Wiebers DO et al. Recommendations for the management of patients with unruptured intracranial aneurysms: a statement for healthcare professionals from the Stroke Council of the American Heart Association. Stroke 2000; 31: 2742–50.

Hankey GJ, Warlow CP, Sellar RJ. Cerebral angiographic risk in mild cerebrovascular disease. Stroke 1990; 21: 209–22.

Vaphiades MS, Cure J, Kline L. Management of intracranial aneurysm causing a third nerve cranial palsy: MRA, CTA or DSA? Semin Ophthalmol 2008; 23: 143–50.

Grandin CB, Mathurin P, Duprez T et al. Diagnosis of intracranial aneurysms: accuracy of MR angiography at 0.5 T. Am J Neuroradiol 1998; 19: 245–52.

Ida M, Kurisu Y, Yamashita M. MR angiography of ruptured aneurysms in acute subarachnoid hemorrhage. Am J Neuroradiol 1997; 18: 1025–32.

Futatsuya R, Seto H, Kamei T et al. Clinical utility of three-dimensional time-of-flight magnetic resonance angiography for the evaluation of intracranial aneurysms. Clin Imaging 1994; 18: 101–6.

Ross JS, Masaryk TJ, Modic MT et al. Intracranial aneurysms: evaluation by MR angiography. Am J Neuroradiol 1990; 11: 449–56.

Horikoshi T, Fukamachi A, Nishi H, Fukasawa I. Detection of intracranial aneurysms by three-dimensional time-of-flight magnetic resonance angiography. Neuroradiology 1994; 36: 203–7.

Gouliamos A, Gotsis E, Vlahos L et al. Magnetic resonance angiography compared to intra-arterial digital subtraction angiography in patients with subarachnoid haemorrhage. Neuroradiology 1992; 35: 46–9.

Huston J, Nichols DA, Luetmer PH et al. Blinded prospective evaluation of sensitivity of MR angiography to known intracranial aneurysms: importance of aneurysm size. Am J Neuroradiol 1994; 15: 1607–14.

White PM, Wardlaw JM. Unruptured intracranial aneurysms. J Neuroradiol 2003; 30: 336–50.

White PM, Teasdale EM, Wardlaw JM, Easton V. Intracranial aneurysms: CT angiography and MR angiography for detection—prospective blinded comparison in large patient cohort. Radiology 2001; 219: 739–49.

Jacobson DM, Trobe JD. The emerging role of magnetic resonance angiography in the management of patients with third cranial nerve palsy. Ame J Ophthalmol 1999; 128: 94–6.

Anderson GB, Findlay JM, Steinke DE, Ashforth R. Experience with computed tomographic angiography for the detection of intracranial aneurysm in the setting of acute subarachnoid hemorrhage. Neurosurgery 1997; 41: 522–8. (missed 6 aneurysms all 3mm or less)

Piotin M, Gailloud P, Bidaut L et al. CT angiography, MR angiography and rotational digital subtraction angiography for volumetric assessment of intracranial aneurysms. An experimental study. Neuroradiology 2003; 45(6): 404–9.

Chappell ET, Moure FC, Good MC. Comparison of computed tomographic angiography with digital subtraction angiography in the diagnosis of cerebral aneurysms: a meta-analysis. Neurosurgery 2003; 52: 624–31.

Villablanca JP, Jahan R, Hooshi P et al. Detection and characterization of very small cerebral aneurysms by using 2D and 3D helical CT angiography. AJNR Am J Neuroradiol 2002; 23: 1187–98.

Goddard AJ, Tan G, Becker J. Computed tomography angiography for the detection and characterization of intra-cranial aneurysms: current status. Clin Radiol 2005; 60: 1221–36.

CON: A NEGATIVE MRA OR CTA IS ADEQUATE IN THE EVALUATION OF A PUPIL-INVOLVED THIRD NERVE PALSY

Michael S Lee

The pupillary fibers course along the superior aspect of the oculomotor nerve. After leaving the brainstem, the third nerve runs inferior to the posterior communicating (PComm) artery. Aneurysms at the junction of the PComm and the internal carotid arteries enlarge inferiorly and therefore can cause pupil involved third nerve palsies. Unfortunately, microvascular ischemic oculomotor palsies can also cause an efferent pupillary defect in up to 25% of cases.

Aneurysms account for up to 30% of isolated 3rd nerve palsies. The majority of aneurysms (95%) causing third nerve palsies measure at least 5 mm. The smallest aneurysm causing a third nerve palsy that I have found in the English literature is 3 mm. Generally speaking, MRA can reliably detect aneurysms of 5 mm or more and CTA can reliably detect aneurysms of 3 mm or more. Aneurysms as small as 1–2 mm can be detected using CTA but results can be variable depending on the radiologist, software used, and number of detectors on the scanner. However, with third nerve palsies, the localization of a potential aneurysm is well known and it is extremely unlikely that CTA will miss a PComm aneurysm large enough to cause a third nerve palsy. As technology continues to improve, the increasing prevalence of 128- to 256- detector CTA will only increase the resolution of this modality. Large studies of cerebral aneurysms causing isolated third nerve palsies have found outstanding detection rates with CTA compared to conventional catheter angiography. Mathew and colleagues (1) recently studied 137 patients with isolated third nerve palsy. Multidetector CTA identified a causative aneurysm in 27 patients and four incidental aneurysms. Catheter angiography did not detect any other aneurysms among these patients. Of the remaining 110 patients with a normal CTA, none developed evidence of an aneurysm such as aberrant regeneration after a mean of 8 months followup.(1)

Catheter angiography carries a 1–2% risk of neurologic or systemic complication and the risk increases with the age of the patient and the presence of cerebral atherosclerosis. Angiography remains the gold standard for identifying aneurysms; however the risk of the procedure may not outweigh the increase in diagnostic ability among certain patients. The patient described is older and carries a strong vasculopathic medical history. He could very well have a microvascular third nerve palsy with pupillary involvement. I would order an MRI and MRA to rule out both an aneurysm and any other compressive or inflammatory lesion that could cause an oculomotor palsy. Since his pupil is involved there is more concern for an aneurysm. Noninvasive imaging with CTA increases our diagnostic yield for smaller aneurysms. If the CTA were also negative and I felt that I could trust my neuroradiologist, then I would recommend observation. Generally speaking the radiologist knows where to look for an aneurysm and if present, it ought to measure at the very least 3–5 mm in size which is generally acceptable for the CTA. Depending on where you practice, a well-trained neuroradiologist may not be available. In the absence of a good interpretation, this case might warrant a catheter angiogram by an interventionalist in order to comfortably rule out an aneurysm.

REFERENCE

1. Mathew R, Teasdale E, McFadzean RM. Multidetector computed tomographic angiography in isolated third nerve palsy. Ophthalmology 2008; 115: 1411–15.

SUMMARY

Although the techniques of noncatheter angiography have improved and continue to improve the "gold standard" for detection of an aneurysm probably remains catheter angiography. The real questions in third nerve palsy are "What is the "pre-test" likelihood for aneurysm?" in a given patient and "What are the risks in this individual patient for catheter angiography?". The question is complicated by the fact that different institutions have different quality and bias in interpretation for their own angiography options. Although MRA might be better in one institution, CTA might be better at another institution particularly if that particular institution has special techniques or expertise in one technique over another. Most institutions prefer CTA for the task of ruling out aneurysm. In addition, seeing the relevant arterial anatomy in 3-dimensional rotational space with the source images and using specialized software and idealized projection monitors with the neuroradiologist in the room can make the difference between missing and seeing a small aneurysm. It should be obvious that there cannot be a single answer for every patient and that catheter angiogram may still be necessary for patients with a high clinical suspicion for aneurysm even with a completely negative MRA or CTA. The other problem encountered clinically is that the MRI is the superior study for nonaneurysmal causes of third nerve palsy (e.g., tumors) and it is much easier to obtain an MRI and MRA than an MRI with a CTA. At some institutions CTA is the first line study and if negative the clinician could proceed to MRI for the nonaneurysmal etiologies. Catheter angiography has inherent risks and these must be weighed on an individual basis against the risk of missing a potentially life threatening aneurysm. Ultimately the clinician's, neuroradiologist's, and institutional experience, quality of imaging and confidence in the study will be different from place to place.

13 Do erectile dysfunction agents cause anterior ischemic optic neuropathy?

A 54-year-old male is seen in the local ophthalmology office. He has a past medical history significant for hyperlipidemia, hypertension, and erectile dysfunction. He woke yesterday with sudden awareness of decreased vision in his right eye, which has not improved. His visual acuity is 20/50 OD and 20/16 OS. He has a 1.2 log unit RAPD on the right. His visual fields reveal an inferior altitudinal defect on the right, and normal visual field on the left (Figure 13.1 and 13.2). His slit lamp exam is normal. His dilated fundus exam reveals disc edema on the right only, and a crowded optic nerve configuration on the left (Figure 13.3 and 13.4). OCT of the RNFL was also obtained (Figure 13.5) confirming the disc edema. As the

possibility of permanent vision loss is explained to the patient, and his vasculopathic risk factors are being addressed, the patient reminds the ophthalmologist about his taking medication for erectile dysfunction and asks if the drug caused this to happen.

PRO: ERECTILE DYSFUNCTION AGENTS DO CAUSE ANTERIOR ISCHEMIC OPTIC NEUROPATHY

Karl Golnik

Patients with nonarteritic anterior ischemic optic neuropathy (NAION) typically are > 50 years of age and present with

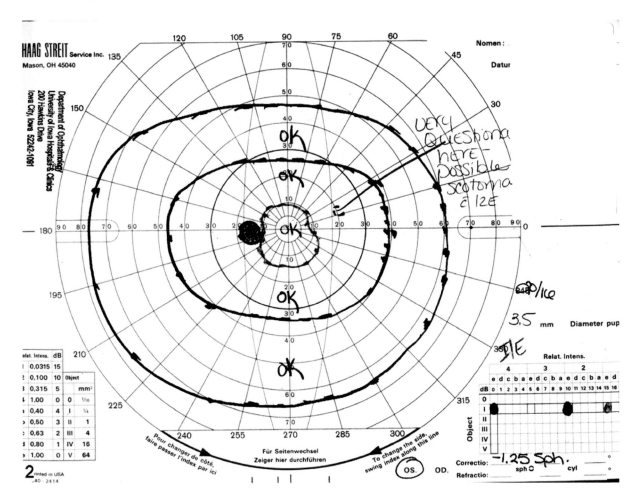

Figure 13.1 Normal visual field, left eye.

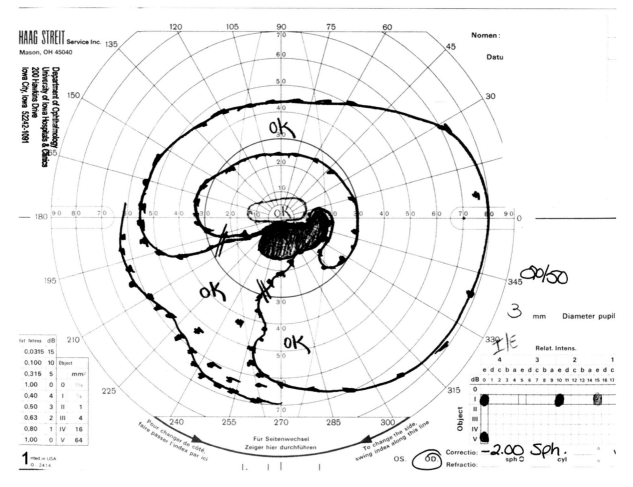

Figure 13.2 Goldmann visual field, right eye, showing an inferior altitudinal defect.

painless, sudden visual loss. Vision improves (three or more lines) in about 43% of patients over 6 months. Reported risk factors and associated conditions include age, hypertension, nocturnal hypotension, diabetes mellitus, cigarette use, hypercholesterolemia, hypertriglyceridemia, elevated fibrinogen, hypercoagulable states, acute blood loss, anemia, elevated intraocular pressure, migraine, sleep apnea, and postcataract surgery. Patients with NAION also almost always have the "disc-at-risk"; a small disc with cup of 0.0–0.2.(1) The annual incidence of NAION is between 2.3 and 10.2 per 100,000 persons over the age of 50.(2, 3)

Recently, NAION has been reported in patients using agents for treatment of erectile dysfunction. sildenafil, (Viagra), vardenafil (Levitra), and longer acting tadalafil (Cialis) are selective inhibitors of cyclic guanosine monophosphate (cGMP) specific phosphodiesterase type 5 (PDE-5). These agents work by enhancing the effect of nitric oxide and cyclic guanosine

monophosphate pathway (GMP) which leads to smooth muscle relaxation in the corpus cavernosum, allowing inflow of blood during sexual stimulation. The most common side effects are headache and facial flushing. A variety of visual side effects have been reported and include changes in color perception, objects have colored tinges (usually blue or bluegreen, may be pink or yellow), decreased color vision, dark colors appear darker, blurred vision, central haze, transitory decreased vision, changes in light perception, increased perception of brightness, flashing lights especially when blinking, Electroretinography (ERG) changes (transient), conjunctival hyperemia, ocular pain, and photophobia.

In 2002 Pomeranz and associates reported five patients (which included two previous single case reports) who developed NAION after sildenafil ingestion. They were men, aged 42, 52, 59, 62, 69 and each noted visual disturbance within 45 minutes to 12 hours of ingestion (in one case the duration

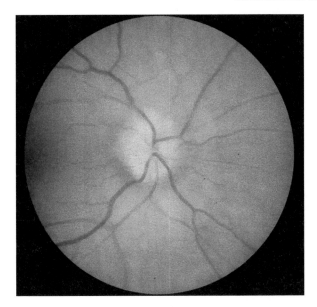 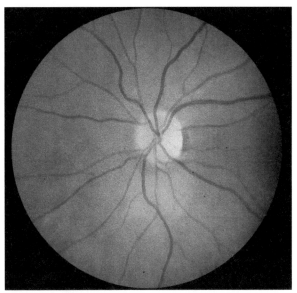

Figure 13.3 Optic nerve, right eye, showing superior segmental disc edema consistent with ischemic optic neuropathy.

Figure 13.4 Optic nerve, left eye, demonstrating the "disc-at-risk" consistent with a diagnosis of nonarteritic ischemic optic neuropathy in the fellow eye.

was unclear). All of the patients had the "disc-at-risk." Other than age, four of the five had no other NAION risk factors. The authors felt the rapid onset of ocular symptoms in four of five subjects is supportive of an association between use of sildenafil and NAION. Bollinger and Lee reported a very interesting 67-year-old man who developed transient inferior visual field loss in the right eye within 2 hours of taking the first four doses of tadalafil. After the fifth dose, he developed NAION in the right eye with persistent inferior visual field loss. Carter reviewed these cases and 16 others and found an age range of 42–69 years. Twelve patients had a small cup:disc ratio, one patient was "normal" and the cup:disc ratio was unknown in eight patients. Six of the 21 patients had first NAION symptoms after what would be the expected length of action of the drug used.

Only one case-controlled study has investigated the possible relationship between these drugs and NAION. McGwin and associates retrospectively studied 38 patients with NAION and 38 age and sex matched control patients. Unmasked phone interviews were done to ascertain medication use. They reported that men with NAION had an odds ratio of 1.75 (95% confidence interval (CI), 0.48–6.30) (statistically not significant) for having used PDE-5 inhibitors. However, a significant odds ratio of 10.7 (95% CI, 1.3–95.8) was seen in those with a history of myocardial infarction and near significant odds ratio of 6.9 (95% CI 0.8–63.6) in men with a history of hypertension. They concluded that for men with a history of myocardial infarction or hypertension the use of Viagra or Cialis may

increase the risk of NAION. It should be noted that this type of study has limitations including the possibility of biases in the selection of controls, biases from nonmasked interviewers, and underreported use of erectile dysfunction drugs by controls to telephone interviewers.

Another issue regarding these medications is the mechanism of action that would result in NAION. The mechanism by which these medications might damage the optic nerve is not as well understood. It has been theorized that sildenafil, which works through the nitric oxide cyclic GMP pathway, may alter the perfusion of the optic nerve head by modifying nitric oxide levels. Theories regarding mechanism of action are complicated by our lack of understanding of the usual mechanism of NAION development.

The Food and Drug Administration has issued a statement regarding reports of patients experiencing a sudden loss of vision attributed to NAION after taking Viagra, Cialis, and Levitra. This statement is clear that no casual link has been established between these medications and the occurrence of NAION. However, the current sidenafil medication label states: "Physicians should advise patients to stop use of all PDE-5 inhibitors, including VIAGRA, and seek medical attention in the event of a sudden loss of vision in one or both eyes. Such an event may be a sign of nonarteritic anterior ischemic optic neuropathy (NAION), a cause of decreased vision including permanent loss of vision, that has been reported rarely postmarketing in temporal association with the use of all PDE-5 inhibitors. It is not possible to determine whether these events are related directly to the use

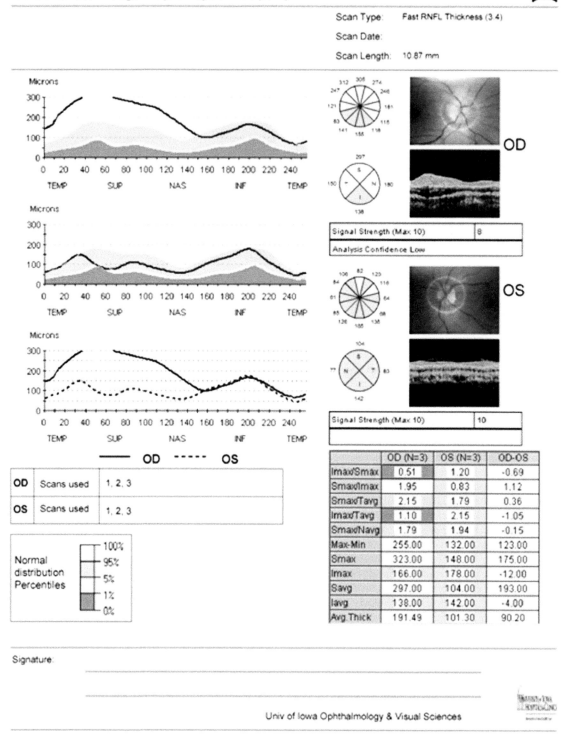

STRATUS OCT
RNFL Thickness Average Analysis Report - 4.0.3 (0072)

Scan Type: Fast RNFL Thickness (3.4)
Scan Date:
Scan Length: 10.87 mm

	OD (N=3)	OS (N=3)	OD-OS
Imax/Smax	0.51	1.20	-0.69
Smax/Imax	1.95	0.83	1.12
Smax/Tavg	2.15	1.79	0.36
Imax/Tavg	1.10	2.15	-1.05
Smax/Navg	1.79	1.94	-0.15
Max-Min	255.00	132.00	123.00
Smax	323.00	148.00	175.00
Imax	166.00	178.00	-12.00
Savg	297.00	104.00	193.00
Iavg	138.00	142.00	-4.00
Avg.Thick	191.49	101.30	90.20

Univ of Iowa Ophthalmology & Visual Sciences

Figure 13.5 OCT of the RNFL, demonstrating the segmental nerve fiber layer swelling.

of PDE-5 inhibitors or to other factors. Physicians should also discuss with patients the increased risk of NAION in individuals who have already experienced NAION in one eye, including whether such individuals could be adversely affected by use of vasodilators, such as PDE-5 inhibitors."

Thus it is unclear whether these medications are risk factors for NAION. Further studies designed to answer this question are being initiated at the present time. Meanwhile, I ask all my patients with NAION if they are using erectile dysfunction medications and if they are, I advise them to stop. I explain to them that there is no definite proof of causation but if it were me, I would discontinue use. I do not recommend stopping these medications in patients I see for non-NAION diagnoses even if they have other risk factors for NAION.

REFERENCES

1. Burde RM. Optic disc risk factors for nonarteritic anterior ischemic optic neuropathy. Am J Ophthalmol 1993; 116: 759–64.
2. Hattenhauer MG, Leavitt JA, Hodge DO, Grill R, Gray DT. Incidence of nonarteritic anterior ischemic optic neuropathy. Am J Ophthalmol 1997; 123: 103–7.
3. Johnson LN, Arnold AC. Incidence of nonarteritic and arteritic anterior ischemic optic neuropathy. Population-based study in the state of Missouri and Los Angeles County, California. J Neuroophthalmol 1994; 14: 38–44.

BIBLIOGRAPHY

Ischemic optic neuropathy decompression trial research group. Optic nerve decompression surgery for non arteritic anterior ischemic optic neuropathy (NAION) is not effective and may be harmful. JAMA 1995; 273: 625–32.

Ischemic optic neuropathy decompression trial research group. Characteristics of patients with non-arteritic anterior ischemic optic neuropathy eligible for the ischemic optic neuropathy decompression trial. Arch Ophthalmol 1996; 114: 1366–74.

Hayreh SS, Joos KM, Podhajsky PA, Long CR. Systemic conditions associated with non-arteritic anterior ischemic optic neuropathy. Am J Ophthalmol 1994; 118: 766–80.

Hayreh SS. Role of nocturnal hypotension in optic nerve head ischemic disorders. Ophthalmologica 1999; 213: 76–96.

Mojon DS, Hedges TR III, Ehrenberg B et al. Association between sleep Apnea syndrome and nonarteritic anterior ischemic optic neuropathy. Arch Ophthalmol 2002; 120: 601–5.

Palombi K, Renard E, Levy P et al. Non-arteritic anterior ischaemic optic neuropathy is nearly systematically associated with obstructive sleep apnoea. Br J Ophthalmol 2006; 90: 879–82.

Talks SJ, Chong NH, Gibson JM, Dodson PM. Fibrinogen, cholesterol and smoking as risk factors for non-arteritic anterior ischaemic optic neuropathy. Eye 1995; 9: 85–8.

Chung SM, Gay CA, McCrary JA 3rd. Nonarteritic ischemic optic neuropathy. The impact of tobacco use. Ophthalmology 1994; 101: 779–82.

Deramo VA, Sergott RC, Augsburger JJ et al. Ischemic optic neuropathy as the first manifestation of elevated cholesterol levels in young patients. Ophthalmology 2003; 110: 1041–5.

Fraunfelder FW. Visual side effects associated with erectile dysfunction agents. Am J Ophthalmol 2005; 140: 723–4.

Pomeranz HD, Smith KH, Hart WM, Egan RA. Sildenafil-associated nonarteritic anterior ischemic optic neuropathy. Ophthalmology 2002; 109: 584–7.

Bollinger K, Lee MS. Recurrent visual field defect and ischemic optic neuropathy associated with tadalafil rechallenge. Arch Ophthalmol 2005; 123: 400–1.

Carter JE. Anterior ischemic optic neuropathy and stroke with use of PDE-5 inhibitors for erectile dysfunction: cause or coincidence? J Neurol Sci 2007; 262: 89–97.

Peter NM, Singh MV, Fox MV. Taladafil-associated anterior ischaemic optic neuropathy. Eye 2005; 19: 715–17.

Pomeranz HD, Bhavasar AR. Nonarteritic ischemic optic neuropathy developing soon after use of sildenafil (Viagra): a report of seven new cases. J Neuroophthalmol 2005; 25: 9–13.

Dheer S, Rekhi GS, Merlyn S. Sidenafil associated anterior ischemic optic neuropathy. J Assoc Physicians India 2002; 50: 265.

Boshier A, Pambakian N, Sharkir SA. A case of nonarteritic anterior ischemic optic neuropathy in a male patient taking sildenafil. Int J Pharmacol Ther 2002; 40: 422–3.

Gruhn N, Fledeliurs HC. Unilateral optic neuropathy associated with sildenafil intake. Acta Ophth Scand 2005; 1004: 131–2.

Sinha S, Pathak-Ray V, Ahluvalia H et al. Viagra or what? Eye 2004; 18: 446–8.

McGwin G, Vaphiades MS, Hall TA, Owsley C. Non-arteritic anterior ischaemic optic neuropathy and the treatment of erectile dysfunction. Br J Ophthalmol 2006; 90: 154–7.

Danesh-Meyer HV, Levin LA. Erectile dysfunction drugs and risk of anterior ischaemic optic neuropathy: casual or causal association? Br J Ophthalmol 2007; 91: 1551–5.

U.S. Food and Drug Administration most recent Viagra label available at http://www.fda.gov/cder/foi/label/2008/020895 s026lbl.pdf accessed July 14, accessed July 14, 2008.

CON: ERECTILE DYSFUNCTION AGENTS HAVE NEVER BEEN PROVEN TO CAUSE ISCHEMIC OPTIC NEUROPATHY

Timothy J McCulley and Michael K Yoon

In the case presented here, of a patient who specifically questioned whether the erectile dysfunction medication caused the NAION, my response would be that it is unlikely. The patient is a classic or "textbook" case of spontaneous NAION. He is a 54-year-old man. He has crowded optic nerves. In addition, he has a past medical history significant for hyperlipidemia and hypertension. These are shared risk factors for both NAION and erectile dysfunction, and the occurrence of NAION in such an individual is very likely coincidental. However, I would inform the patient of the anecdotal evidence and leave the decision as to whether or not to continue the use of the medication for him to make.

Nonarteritic ischemic optic neuropathy (NAION) is the most common optic neuropathy in people older than 50 years of age. Although the specifics remain debated, it is generally believed to represent an ischemic event in the posterior ciliary arteries. Risk factors include conditions which predispose to atherosclerosis such as hypertension, diabetes, hypercholesterolemia, and smoking. Also, the vast majority of NAION events involve patients with a "crowded disc" or small cup-to-disc ratio. There is another subset of diseases which are more commonly encountered in patients with NAION due to shared risk factors, but that do not contribute to the development of NAION. Examples include cardiovascular and coronary artery disease. Lastly there is a collection of abnormalities that probably do contribute to the development of NAION, but only in a minority of patients.

There remain some diseases or other factors, such as medications, for which anecdotal associations have been described and for which a logical explanation for causation might be imagined; however, the nature of the relationship (if any) has yet to be definitively established. Chronic obstructive sleep apnea and conditions leading to hyper-coagulation fall into this category. Similarly, several accounts of NAION occurring in patients that use phosphodiesterase-5 (PDE-5) inhibitors to treat erectile dysfunction, such as Viagra (sildenafil), Cialis (tadalafil) or Levitra (vardenafil), have been reported. Yet no systematic evaluation has illuminated the exact nature of the relationship.

Although the possible association between erectile dysfunction drugs has been considered based on a few case reports, the relationship has not been well-supported by the data. The speculation in the medical community regarding the potential relationship between erectile dysfunction drugs and NAION is based primarily on isolated case reports. Amidst these published cases, most shared the common risk factors of a crowded nerve and conditions predisposing to atherosclerosis. Moreover, the reports of NAION temporally associated with the use of PDE-5 inhibitors like tadalafil have been rare. As of May 2005, the FDA had received 38 case reports involving Viagra (sildenafil) and only four reports of NAION with Cialis (tadalafil). This is in the context of millions of prescriptions of the erectile dysfunction medications. Case reports are only anecdotal and are not accepted as sufficient to establish a causal relationship. In this setting, the validity of these case reports is particularly problematic. Men who use PDE-5 inhibitors for erectile dysfunction are not a random sample. Many of the risk factors for erectile dysfunction are also risk factors for NAION and one would expect that some men being treated for erectile dysfunction would coincidentally experience NAION.

There have been three "re-challenge" cases reported. In 2005, Bollinger and Lee described a patient with transient followed by permanent visual field loss with successive sildenafil administration. In the same year, Pomeranz and Bhavsar described a patient with bilateral sequential AION following repeat use of sildenafil. Most recently, Pepin and Pitha-Rowe reported a patient with "stepwise decline in visual field" with continued sildenafil use. Although admittedly suggestive, even these cases fail to confirm a causative relationship between sildenafil and NAION. Bilateral sequential, recurrent, and progressive (stepwise or continuous) NAION have all been described in patients not taking erectile dysfunction medications and even these cases are likely coincidental to their use.

There are no postmarketing randomized controlled trials (the gold standard in epidemiology) evaluating the use of PDE-5 inhibitors and NAION. One retrospective case-control study looking at NAION and the use of these drugs has been published. Dr. Gerald McGwin and researchers at the University of Alabama at Birmingham conducted a retrospective matched case-control study of NAION and use of erectile dysfunction drugs. That study found that overall males with NAION were no more likely to report a history of use of erectile dysfunction medication than a similarly aged control group. Using the electronic National Veterans Health Administration's pharmacy and clinical databases, Margo and French looked at the occurrence of NAION in patients prescribed use of erectile dysfunction medications. They found that there was slight increase in incidence of NAION among men over 50 years of age who had been prescribed PDE-5 inhibitors compared to those who had not, relative risk 1.10 (95% 1.02–1.20). This could reflect a causal relationship but, as acknowledged by the authors, is likely explained simply by erectile dysfunction and NAION sharing the same vascular risk factors.

Thus, it is my opinion that the existing scientific data are insufficient to establish a causal relationship between PDE-5 inhibitors and NAION. In managing patients with NAION, I think it is appropriate and reasonable to inform them of a possible relationship. I would suggest having and documenting this discussion with all patients with NAION and to inform your patients of a potential relationship but also for medicolegal reasons. Conversely, I think it is inappropriate to overstate the threat of erectile dysfunction drugs, which may be your patients' only effective remedy for their erectile dysfunction. On an individual basis, I leave it to my patients to weigh the benefit vs. the theoretical risk of erectile dysfunction drugs and let them make an informed decision of whether or not to continue their use.

BIBLIOGRAPHY

Bollinger K, Lee MS. Recurrent visual field defect and ischemic optic neuropathy associated with tadalafil rechallenge. Arch Ophthalmol 2005; 125: 400–1.

Pomeranz HD, Bhavsar AR. Nonarteritic ischemic optic neuropathy developing soon after use of sildenafil (Viagra):

a report of seven new cases. J Neuroophthalmol 2005; 25: 9–13.

Pepin S, Pitha-Rowe I. Stepwise decline in visual field after serial sildenafil use. J Neuroophthalmol 2008; 28: 76–7.

Akash R, Hrishikesk D, Amith P et al. Case report: association of combined nonarteritic anterior ischemic optic neuropathy (NAION) and obstruction of cilioretinal artery with overdose of Viagra. J Ocul Pharmacol Ther 2005; 21: 315–7.

Cunningham AV, Smith KH. Anterior ischemic optic neuropathy associated with Viagra. J Neuroophthalmol 2001; 21: 22–5.

Pomeranz HD, Smith KH, Hart WM Jr et al. Sildenafil-associated nonarteritic anterior ischemic optic neuropathy. Ophthalmology 2002; 109: 584–7.

Escaravage GK Jr, Wright JD Jr, Givre SJ. Tadalafil associated with anterior ischemic optic neuropathy. Arch Ophthalmol 2005; 123: 399–400.

Hatzichristou D. Phosphodiesterase 5 Inhibitors and non-arteritic anterior ischemic optic neuropathy (NAION): coincidence or causality? J Sex Med 2005; 2: 751–58.

McGwin G Jr, Vaphiades MS, Hall TA, Owsley C. Non-arteritic anterior ischaemic optic neuropathy and the treatment of erectile dysfunction. Br J Ophthalmol 2006; 90: 154–7.

Margo CE, French DD. Ischemic optic neuropathy in male veterans prescribed phosphodiesterase-5 inhibitors. Am J Ophthalmol 2007; 143: 538–9.

SUMMARY

The erectile dysfunction agents are "big business" and establishing or refuting a causal relationship between these agents and nonarteritic anterior ischemic optic neuropathy (NAION) is a "high stakes" issue. There is no doubt that there have been well documented reports of visual loss associated with these agents. There is a biologically plausible mechanism (hypotension) for NAION, close temporal relationship in some cases between taking the drug and visual loss and even a few re-challenge cases. The strongest cases for causality are the bilateral simultaneous or rapidly sequential rechallenge cases occurring close to dosing. The weakest cases are unilateral, non-rechallenged cases occurring far from the exposure dose. On the other hand there have been literally billions of doses prescribed, many of the cases have onset of visual loss that is not coherent with the known pharmacokinetics (i.e., peak onset, half life) for the drug and the number of cases remains low despite millions of prescriptions worldwide. A proper case-control study with sufficient sample size and statistical power remains to be published but the pharmaceutical companies are under pressure from the Food and Drug Administration (FDA) to provide evidence for a lack of association for the drugs. Stay tuned to this channel for the outcome but at this point the causal relationship for the erectile dysfunction agents and NAION has not been proven.

14   Does amiodarone produce an optic neuropathy?

An 84-year-old man presents with a past medical history of abdominal aortic aneurysm, diabetes type II, hypertension, and hypercholesterolemia. He underwent a coronary artery bypass 5 months ago, and during the immediate postoperative period, he experienced atrial fibrillation for which he was started on amiodarone. His other medications are aspirin, atorvastatin, fosinopril, glipizide, and carvedilol twice daily, including one dose at bedtime. He described a progressive decrease in his vision OU over 3 months and additional worsening 3 weeks ago in his left eye. He denies any symptom suggestive of temporal arteritis and his erythrocyte sedimentation rate is normal. On examination, his visual acuity is 20/20 OD and 20/30 OS. There is a 0.9 log unit RAPD OS. Dilated fundus exam revealed marked asteroid hyalosis OD and optic disc edema OU, shown in Figures 14.1 and 14.2, Goldman perimetry shows visual field loss OU, and illustrated in Figures 14.3 and 14.4.

PRO: AMIODARONE DOES CAUSE AN OPTIC NEUROPATHY

Eric Eggenberger
Several agents, including medications, have been associated with optic neuropathy. When evaluating the relationship

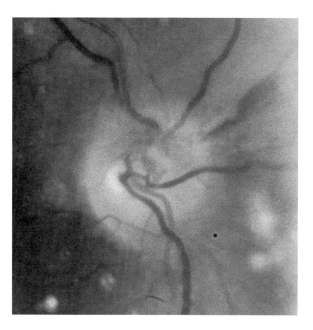

Figure 14.1 Optic nerve photograph, right eye, showing early optic disc edema with peripapillary hemorrhages.

Table Koch's Postulates for Causal Association.

1. Time order
2. Biologic plausibility
3. Dose effect
4. Rechallenge
5. Reproducibility

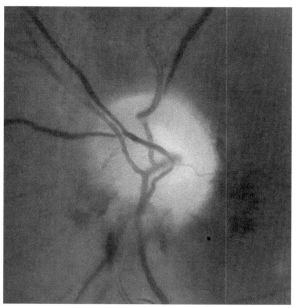

Figure 14.2 Optic nerve photograph, left eye, also showing early optic disc edema and numerous hemorrhages surrounding the optic nerve.

between exposure and an effect, it is important to bear in mind the difference between association and causation. The assignment of a causal association is often based on fulfillment of modified Koch's postulates (Table).

Although several variations of optic neuropathy (unilateral or bilateral, anterior or retrobulbar) have been associated with amiodarone use, the bilateral simultaneous anterior optic neuropathy variant contains the most convincing causal relationship in accord with Koch's postulates; this is the group 1 ("probable amiodarone optic neuropathy") simultaneous bilateral disc edema subcohort described by Purvin et al.(2) The unilateral anterior optic neuropathy cases with small cup-to-disc ratio on the fellow eye most logically represent ischemic pathophysiology in a population with

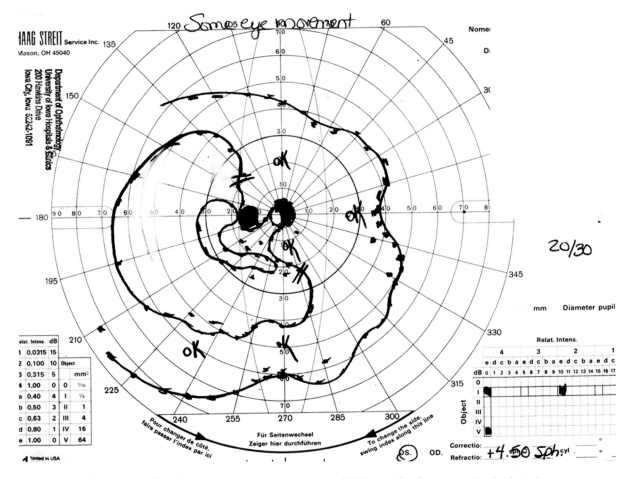

Figure 14.3 Goldmann visual field, left eye, showing mostly superior field loss, with relative central and inferior loss.

prevalent vascular risk factors; re-classification of such cases into the amiodarone causal category would require more definitive means to ascribe pathophysiology than are currently available.

The time order constraint is the easiest to satisfy; most convincing cases of amiodarone related optic neuropathy have occurred weeks to months following initiation of the medication. The half-life of amiodarone is nearly 2 months, so exposure remains long after dosing is ceased. Biologic plausibility of amiodarone optic neuropathy is provided by histopathologic evaluation of the optic nerve in an amiodarone-treated patient without visual symptoms; multiple lamellar inclusion bodies within large axons without axonal loss were observed. Amiodarone-treated mice have shown similar inclusions in glial cells. It is possible that intraneuronal amiodarone accumulation leads directly to axonal swelling, or that glial cell deposition produces swelling of these cells leading to axonal transport obstruction. The dose effect requisite is addressed in the few cases of amiodarone-related optic neuropathy in which dose reduction has been followed by stabilization of the process. Re-challenge has not been reported with amiodarone, likely because of the risk of further visual loss. Reproducibility has been effectively satisfied by the number of cases reported.

Although some cases of anterior ischemic optic neuropathy will be expected to occur in the target population requiring agents such as amiodarone, we agree with Purvin et al.(2) and others that bilateral and simultaneous cases with prolonged disc edema are sufficiently distinct to warrant classification as amiodarone-related. Macaluso and colleagues reviewed data from 73 patients with reported amiodarone-related optic neuropathy, and highlighted distinguishing features of the condition. These authors highlighted the insidious onset, slow progression, bilateral visual loss, and prolonged disk edema that stabilized within several months of discontinuing the medication; in contrast, nonarteritic anterior ischemic optic neuropathy typically presents with acute, unilateral visual loss that is usually complete at onset, and resolution of disk edema over several weeks.

Sreih et al.(1) observed 3 cases of amiodarone-associated anterior optic neuropathy within a single electrophysiology practice,

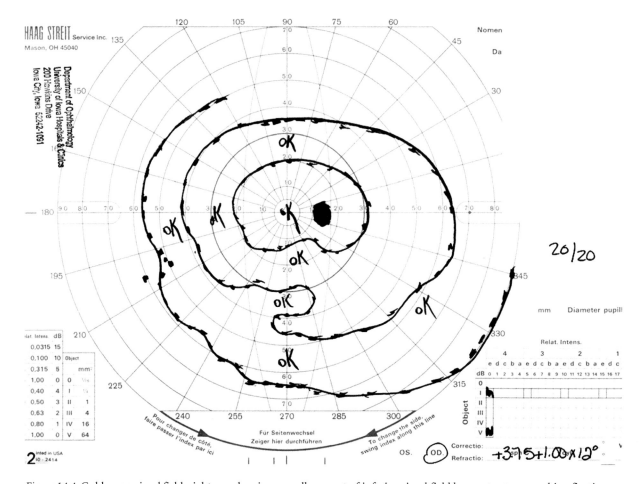

Figure 14.4 Goldmann visual field, right eye, showing a small amount of inferior visual field loss, not yet approaching fixation.

and recommended routine ophthalmology monitoring during such therapy; we are unsure that routine monitoring has the requisite evidence base to justify its usefulness, sensitivity, and cost.

Thus, it is my opinion that unilateral cases of anterior optic neuropathy in patients with typical risk factors for ischemic events likely represents chance association, while bilateral simultaneous anterior optic neuropathy with prolonged disc edema likely represents a distinct medication-related toxic optic neuropathy. In such cases, the risk of discontinuing the medication must be weighed against the risk of continued medication use, bearing in mind that many of these patients require such medications for life-threatening cardiac conditions.

REFERENCES

1. Sreih AG, Schoenfeld MH, Marieb MA. Optic neuropathy following amiodarone therapy. Pacing Clin Electrophysiol 2006; 22(7): 1108–10.
2. Purvin V, Kawasaki A, Borruat FX. Optic neuropathy in patients using amiodarone. Arch Ophthalmol 2006; 124: 696–701.

BIBLIOGRAPHY

Macaluso DC, Shults WT, Fraunfelder FT. Features of amiodarone-induced optic neuropathy. Am J Ophthalmol 1999; 127(5): 610–2.

Mansour AM, Puklin JE, O'Grady R. Optic nerve ultrastructure following amiodarone therapy. J Clin Neuroophthalmol 1988; 8: 231–7.

Costa-Jussa FR, Jacobs JM. The pathology of amiodarone neurotoxicity, I: experimental studies with reference to changes in other tissues. Brain 1985; 108: 735–52.

CON: AMIODARONE DOES NOT CAUSE AN OPTIC NEUROPATHY

Timothy J McCulley and Shelley Day

Several studies have attempted to differentiate the characteristics of supposed amiodarone-related optic neuropathy *versus* NAION. Amiodarone-related optic neuropathy is felt to be more likely to cause insidious onset of visual loss, to involve both eyes simultaneously, and to cause protracted disc edema over months. Hayreh suggested, "most importantly, the clinical features of the optic neuropathy in patients taking amiodarone are typical of NAION rather than a toxic optic neuropathy," and most likely represent a variation along the spectrum of NAION rather than an entirely separate entity. In fact many argue that "amiodarone-related optic neuropathy" is not a distinct entity at all, but rather a coincidental occurrence of spontaneous NAION in at risk patients who are the same individual in whom amiodarone is likely to be prescribed.

Since the first reports of presumed "amiodarone-related optic neuropathy" in 1987, numerous additional cases have been published but no subsequent study has established a causal link between amiodarone use and optic neuropathy. The majority of patients who take amiodarone as antiarrhythmic therapy have the same vascular risk factors as patients who develop nonarteritic ischemic optic neuropathy (NAION) and it would be expected that AION would be encountered in this same group of patients. The concern of an association between AION and amiodarone has been fueled by the medico-legal fears. Despite the absence of an established causal effect of amiodarone in optic neuropathy, Wyeth Ayerst, the manufacturer of amiodarone, was fined in excess of $20 million for failure to include blindness as a potential adverse effect of this medication. This is just one of many examples where the threat of legal consequence is the basis for medical decisions, often at the unfortunate expense of patient care. The following outlines the evidence or rather the lack of support for a causal relationship between amiodarone and an optic neuropathy.

Fraunfelder and Shults recently described 7 criteria, which must be met to establish a causal relationship between a medication and a medical condition: 1) temporal association, 2) dose-response relationship, 3) positive de-challenge evidence, 4) positive re-challenge evidence, 5) a plausible causal mechanism for the agent, 6) a "class effect," and 7) lack of a plausible alternative explanation. As outlined below, none of these criteria have been satisfactorily fulfilled for amiodarone-associated optic neuropathy.

1) Is the rate of optic neuropathy higher in patients who take amiodarone?

An increased incidence has yet to be adequately demonstrated. In their initial 1987 publication, Feiner et al.(1) reported the rate of optic neuropathy in 447 Mayo Clinic patients taking amiodarone at 1.8%. This is comparatively higher than the previously published epidemiologic study which reported a 0.3% annual incidence of NAION in the general population over the age of 50. Although suggestive, these patient populations are not comparable. In a recent editorial Younge stated that "the higher incidence of AION in amiodarone users than in an age-matched group of nonusers…could have been due to chance alone, because most of the amiodarone users were likely to have had substantially more vascular disease." To date, no case-control study has been undertaken to examine the incidence of NAION in patients with multiple cardiovascular risk factors who do not take amiodarone compared with those who do.

2) Is there a temporal association between use of amiodarone and development of optic neuropathy?

Uniformity in the timing between initiation of amiodarone therapy and AION onset has not been seen. Feiner et al.(1) described a very broad range of time from initiation of amiodarone therapy to development of optic neuropathy, from 1 to 72 months with a mean of 10.6 months. Another series of 22 patients reported an interval between initiation of amiodarone to onset of visual symptoms of anywhere from 1 to 22 months (mean, 6 months). The wide range in the time to development of optic neuropathy is all the more surprising given that amiodarone is often given in a higher loading dose (such as 1,600 mg/day for 1–3 weeks for ventricular arrhythmias) at the initiation of therapy.

3) Does optic disk swelling, visual acuity, or visual field defects improve after cessation of the drug?

One of the difficulties in assessing the benefit of discontinuing amiodarone is that spontaneous NAION can also improve without therapy. Given that we would expect both NAION and supposed amiodarone-induced optic neuropathy to improve spontaneously, one measure of the effect of amiodarone is whether these patients take longer to improve if the drug is continued. In the first report of amiodarone associated optic neuropathy, 1 of 13 patients actually experienced worsening of visual acuity after cessation of amiodarone. Gittinger and Asdourian described the disc swelling in 3 eyes of 2 patients on amiodarone, one patient had complete resolution of disc swelling at 1 month despite continuing amiodarone and the other patient had improvement in both discs at a continued but reduced dose of amiodarone Purvin et al.(2) described 22 patients and 3 patients who continued amiodarone therapy despite development of optic neuropathy actually had a faster average time to resolution of disc edema compared to those patients whose amiodarone was discontinued (6 weeks to 3 months *vs.* 6 weeks to > 6 months).

4) Does the optic neuropathy reoccur with re-challenge?

To our knowledge to date, no patient has been reported in the medical literature with an optic neuropathy that recurred with the reinstitution of amiodarone therapy.

5) Is there a dose response?

For supraventricular and ventricular arrhythmias, amiodarone is given in doses ranging from 200 mg/day for maintenance therapy to 1,600 mg/day during the loading period.(7) No correlation between dosage of amiodarone and severity of visual acuity loss, visual field loss, or disk swelling has been made. Admittedly, even if present this would be difficult to establish given the small numbers of patients reported in available case series.

6) Is there a plausible explanation for amiodarone-induced optic neuropathy?

It is well established that amiodarone has ocular side effects, most commonly an amiodarone-induced keratopathy which occurs in >90% of patients. The only histopathologic study of the optic nerve of a patient on a 600 mg/day dose of amiodarone who happened to require enucleation for a choroidal melanoma showed selective accumulation of intracytoplasmic lamellar inclusions in the large axons. The authors inferred that a likely mechanism of optic nerve damage in amiodarone-related optic neuropathy might be a primary lipidosis which may mechanically or biochemically decrease axoplasmic flow. However, this study was of an asymptomatic patient on amiodarone, not a patient with an optic neuropathy, and these optic nerve histopathologic findings have not been confirmed in any other studies.

7) Is there a "class effect"?

Amiodarone is a Class III antiarrhythmic that works primarily by blockage of potassium channels, thereby prolonging repolarization, action potential duration, and the refractory period. Other drugs in this class include sotalol, ibutilide, dofetilide, and azimilide. To date, there have been no published case reports of optic neuropathy associated with any of the other class III antiarrhythmic.

Taken together, a causal relationship between amiodarone and optic neuropathy is far from established. Most cases of amiodarone-associated optic neuropathy occur in patients with multiple cardiovascular risk factors and with much higher risk for spontaneous NAION. There is no clear scientific mechanism or plausible explanation for amiodarone-associated optic neuropathy beyond one histopathologic study showing intracytoplasmic lamellar axonal inclusions. A dose response effect has not been verified, the temporal association of amiodarone use and optic neuropathy varies widely, and the optic disk swelling often resolves despite continued amiodarone use, in cases at an equivalent or faster rate than those patients in whom amiodarone is discontinued. No other Class III antiarrhythmic medications have been associated with a similar optic neuropathy. As Fraunfelder and Shults stated, "the evidence supporting the benefit of amiodarone is far more solid than the evidence of its causing NAION."

Since amiodarone is used in the treatment of life-threatening ventricular and supraventricular arrhythmias, discontinuation of amiodarone should not be undertaken lightly. With regards to amiodarone, until a causal relationship is confirmed, physicians have a responsibility to their patient not to base decisions on medicolegal concerns alone. A reasonable compromise is to inform patients of a possible relationship and consideration for any alternative medications if available. In the patient in question, alternate medications for the management of atrial fibrillation are available and it would not be unreasonable to in this case to consider switching to an alternate medication. However, when amiodarone plays a pivotal role without a close second, the risk of discontinuation might very well exceed the risk of associated visual loss.

REFERENCES

1. Feiner LA Younge BR, Kazmier FJ et al. Optic neuropathy and amiodarone therapy. Mayo Clin Proc 1987; 62: 702–17.
2. Purvin V, Kawasaki A, Borruat FX. Optic neuropathy in patients using amiodarone. Arch Ophthalmol 2006; 124: 696–701.

BIBLIOGRAPHY

Gittinger JW Jr, Asdourian GK. Papillopathy caused by amiodarone. Arch Ophthalmol 1987; 105: 349–51.
Murphy MA, Murphy JF. Amiodarone and optic neuropathy: the heart of the matter. J Neuroophthalmol 2005; 25: 232–6.
Fraunfelder FW, Shults T. Non-arteritic anterior ischemic optic neuropathy, erectile dysfunction drugs, and amiodarone: is there a relationship? J Neuroophthalmol 2006; 26: 1–3.
Younge BR. Amiodarone and ischemic optic neuropathy. J Neuroophthalmol 2007; 27: 85–6.
Vassallo P, Trohman RG. Prescribing amiodarone: an evidence-based review of clinical indications. JAMA 2007; 298: 1312–22.
Mansour AM, Puklin JE, O'Grady R. Optic nerve ultrastructure following amiodarone therapy. J Clin Neuroophthalmol 1988; 8: 231–7.
Macaluso DC, Shults WT, Fraunfelder FT. Features of amiodarone-induced optic neuropathy. Am J Ophthalmol 1999; 127: 610–2.
Hayreh SS. Amiodarone, erectile dysfunction drugs, and non-arteritic ischemic optic neuropathy. J Neuroophthalmol 2006;26: 154–5.

SUMMARY
Amiodarone optic neuropathy (as with the erectile dysfunction agents and NAION) remains a controversial topic. Unfortunately, the same risk factors for NAION are the very reason that a patient is taking amiodarone in the first place. The strongest cases are those that do not look like typical NAION (e.g., prolonged optic disc edema, bilateral and simultaneous onset, and recovery after dechallenge). For medical and medicolegal reasons it might be prudent for the consulting ophthalmologist who diagnoses NAION in a patient on amiodarone to inform the patient and contact the prescribing physician so that an appropriate risk to benefit decision can be made for continuing or discontinuing the drug. There is no doubt that amiodarone has saved many lives in patients with cardiac indications for treatment. Other patients who might be able to switch to an alternative agent especially those with strong clinical suspicion for amiodarone optic neuropathy (i.e., bilateral simultaneous NAION) should be given the option of discontinuing the medication. These patients should be seen in follow up to insure resolution of the disc edema and evaluated for alternative etiologies as well. We recommend photographic documentation of the discs and serial visual field examinations to insure stability and document the course of the optic neuropathy.

15   Should I start my patient with myasthenia gravis on steroids to reduce the chance of generalized myasthenia gravis?

A 47-year-old male presents to the local ophthalmology office complaining of diplopia. It is worse in the evening, and gets worse with reading. The left eyelid is droopy later in the day as well. On examination the visual acuity is 20/20 in each eye. External photograph and motility are shown below (Figures 15.1 and 15.2). He has significant ptosis of the left upper lid, and mild ptosis on the right. He has a Cogan's lid twitch sign. He is given a Fresnel prism which alleviates the diplopia. Prostigmine testing was positive. Serum antiacetylcholine receptor antibodies come back and are all strongly positive (blocking, binding, modifying) suggesting a diagnosis of myasthenia gravis. When he is called back to discuss the results of the testing, he asks if any other parts of the body can be affected by this disease in the same way as his eye movements and eyelids and if anything can be done to prevent generalized disease

PRO: STEROIDS MAY PREVENT GENERALIZED MYASTHENIA GRAVIS IN PATIENTS PRESENTING WITH AN ISOLATED OCULAR FORM OF THE DISEASE

Nicholas Volpe

The ophthalmologist and neuroophthalmologist are frequently on the front lines of the diagnosis of myasthenia gravis. A significant percentage of patients with myasthenia gravis will present with isolated ocular symptoms and these patients will present to the neuroophthalmologist with complaints of either ptosis and/or double vision. Between 30 and 50% of patients with ocular myasthenia gravis (OMG) will go on to develop generalized disease, most with 2 years of onset of ocular symptoms. Like the patient presented, the diagnosis of ocular myasthenia is easily made on the clinical exam based on the presence of variable, fatigueable ptosis, with Cogan's eyelid twitch and/or orbicularis weakness and is further supported by variable pattern of eye movement abnormalities, which changes and/or fatigues during the course of the exam.

Once a diagnosis has been confirmed by ancillary testing such as ice testing, Tensilon/Prostigmin testing, acetylcholine receptor antibodies or EMG, then the neuroophthalmologist and/or neurologist must consider the options for treating the patient. These patients should be investigated for thymoma with a CT scan or MRI scan of the chest. There is evidence to suggest that treatment of thymoma could eliminate the patient's disease and/or reduce their risk of developing generalized myasthenia.

The mainstay of treatment for myasthenia, both ocular and generalized, includes the use of drugs like pyridostigmine, which works by increasing the available neurotransmitter in the synapse and reducing the fatigue of the muscle by providing more

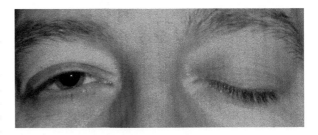

Figure 15.1 Photo showing severe ptosis on the left, and moderate ptosis on the right.

available neurotransmitter. This has been shown to have variable benefits in patients with ocular myasthenia. The second line of treatment involves eliminating and/or reducing the production of the antibodies that are responsible for causing the disease. This reduction in antibody production is accomplished through the use of immunosuppressive agents with the first line of such treatment, the use of oral corticosteroids. Oral corticosteroids are effective in the management of ocular myasthenia improving both ptosis and motility deficits. Once the commitment has been made to treat the patient's symptoms, then the use of oral steroids can also be considered to be beneficial, in reducing the patient's risk of developing generalized myasthenia. The studies that support this as reported by Kupersmith et al.(1, 2), Monsula et al.(3) and Sommer et al.(4), are small, noncontrolled series, with significant potential for selection bias. They suggest in a fairly compelling fashion that there is significant reduction (as low as 10–15% conversion to generalized MG rate) compared to a 50% risk of developing generalized myasthenia in patients not treated with steroids for isolated ocular myasthenia. If the clinician is successful in reducing this risk, then quality of life and long-term prognosis is greatly improved. The mechanism by which low dose oral steroids reduce the risk of generalization is unclear, but would include the possibilities of altering the immune production of the antibody and/or just reducing antibody production to the point where generalized symptoms are less likely to develop. This potential efficacy needs to be carefully weighed against the long term risk of corticosteroid treatment.

Thus, because oral steroids clearly help a significant percentage of patients with ocular myasthenia manage their symptoms they should be offered to patients with isolated ocular myasthenia and there is now some evidence to suggest that this treatment may reduce their risk of developing generalized myasthenia. It would seem that a prospective clinical trial to address this question is warranted. It would be more controversial, should the

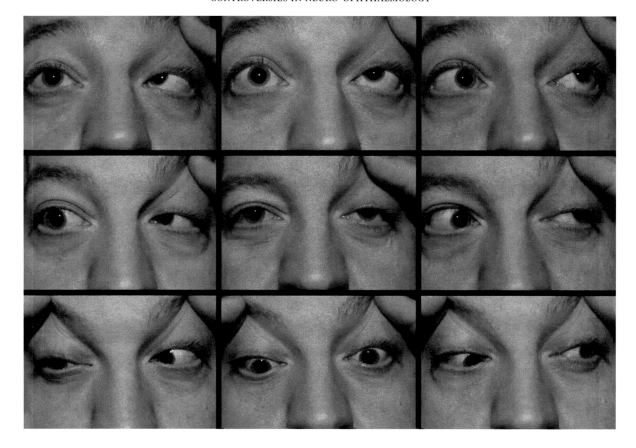

Figure 15.2 Motility photos showing abduction and elevation deficit on the right. The lids are supported due to the ptosis shown in Figure 15.1.

patient not be seeking treatment for their ocular myasthenia. That is if the patient had only mild ptosis, that was not bothering them and/or double vision that only occurred at the very end of the day, deciding to treat this patient with steroids just to reduce the risk of generalization would be difficult. This decision would have to be left to the clinician and patient to make based on the relative risks and benefits of the steroid treatment.

REFERENCES

1. Kupersmith MJ, Latkany R, Homel P. Development of generalized disease at 2 years in patients with ocular myasthenia gravis. Arch Neurol 2003; 60(2): 243–8.
2. Kupersmith MJ, Moster M, Bhuiyan S, Warren F, Weinberg H. Beneficial effects of corticosteroids on ocular myasthenia gravis. Arch Neurol 1996; 53(8): 802–4.
3. Monsula NT, Patwab HS, Knorrb AM, Lesserab RL, Goldsteinb J M. The effect of prednisone on the progression from ocular to generalized myasthenia gravis. J Neurol Sci 2004; 217(2): 131–3.
4. Sommer N, Sigg B, Melms A et al. Ocular myasthenia gravis: response to long-term immunosuppressive treatment. J Neurol Neurosurg Psychiatry 1997; 62(2): 156–62.

BIBLIOGRAPHY

Bever CT Jr, Aquino AV, Penn AS et al. Prognosis of ocular myasthenia. Ann Neurol 1983; 14(5): 516–9.
Robertson NP, Deans J, Compston DA. Myasthenia gravis: a population based epidemiological study in Cambridgeshire, England. J Neurol Neurosurg Psychiatry 1998; 65(4): 492–6.
Shrager JB, Deeb M E, Mick R et al. Transcervical thymectomy for myasthenia gravis achieves results comparable to thymectomy by sternotomy. Ann Thorac Surg 2002; 74(2): 320–6.
Kupersmith MJ, Ying G. Ocular motor dysfunction and ptosis in ocular myasthenia gravis: effects of treatment. Br J Ophthalmol 2005; 89(10): 1330–4.
Kupersmith MJ. Does early treatment of ocular myasthenia gravis with prednisone reduce progression to generalized disease? J Neurol Sci 2004; 217(2): 123–4.

CON: STEROID SHOULD NOT BE GIVEN TO PREVENT ONSET OF GENERALIZED MYASTHENIA GRAVIS

Michael S Lee

Approximately 50% of patients with ocular myasthenia gravis (OMG) will develop generalized myasthenia gravis (GMG) within 2 years of symptom onset and the majority of patients that progress to GMG do so within the first year. The other half of OMG patients will either remain ocular only or enjoy spontaneous remission. A few retrospective studies have suggested that oral corticosteroids may reduce the risk of disease generalization compared to natural history. The Quality Standards Subcommittee of the American Academy of Neurology systematically studied the literature on the medical treatment of OMG. They found five studies that investigated the use of corticosteroids. Two were deemed inadequate in terms of confounding factors and followup. The other three included a total of 118 patients with OMG. These studies did not specifically identify duration of symptoms before starting corticosteroids. They may have included patients with OMG that began steroid treatment 1–2 years after symptom onset, which may skew the data. Conceivably, these patients would not have developed GMG with or without the use of corticosteroids. It is also unclear which, if any, risk factors predict conversion to GMG. Perhaps a factor that predicts conversion led physicians to avoid corticosteroids. These studies have to be viewed with caution since they did not include large numbers of OMG patients or randomization. The possibility that early treatment with corticosteroids may reduce the risk of progression from OMG to GMG is interesting but not entirely logical. The existing literature does not clearly support its widespread use.

As with any disease, the benefit of therapy must outweigh the potential risks. Ideally this involves the least harmful intervention that also alleviates symptoms. If corticosteroids definitively reduced the risk of GMG among patients with OMG then we have a strong argument for its use in OMG. Side effects from short-term exposure of corticosteroids are typically transient and acceptable, but they can affect nearly every system. The more common side effects consist of acne, proximal myopathy, hypertension, hyperphagia, weight gain, peptic ulcer disease, and hyperglycemia. Psychologic effects include insomnia, poor concentration, depression, anxiety, irritability, and psychosis. The studies of corticosteroids in OMG patients have all utilized long-term treatment. Corticosteroid treatment for months to years carries more serious potential complications such as diabetes, glaucoma, cataracts, osteoporosis, aseptic bone necrosis, obesity, and immunosuppression. There have been many calls for a prospective, randomized study to determine this issue. In the absence of definitive efficacy, the risk-benefit ratio of prolonged treatment does not support the use of corticosteroids for OMG to reduce the generalization of the disease.

SUMMARY

Although there has been a lot of anecdotal and retrospective data suggesting that corticosteroids reduce the chance for generalized myasthenia gravis this remains unproven. In the absence of a prospective clinical trial with well defined endpoints and adequate statistical power and sample size the decision to give steroids in ocular myasthenia remains a practice option rather than an evidence-based recommendation. There are other reasons to consider corticosteroids in ocular myasthenia gravis however including the anecdotal superiority of steroids to pyridostigmine in ocular *versus* generalized myasthenia. In addition, steroids have significant and potentially life threatening side effects and an appropriate risk to benefit assessment, coordination of care between the ophthalmologist, neurologist, and primary care, and a frank discussion with the patient should be considered before starting steroids in myasthenia.

16 Does radiation therapy work for thyroid ophthalmopathy?

A 49-year-old woman with a history of hyperthyroidism was treated with radioactive iodine 3 months ago. She is now on hormone replacement and her thyroid function is stable. She noticed progressive periorbital swelling, conjunctival redness, and double vision since 1 month. There is no variability in the diplopia. She stated that her vision is unchanged. Her visual acuity is 20/20 OU and no relative afferent pupillary defect was seen. She has an esotropia of 35 prism diopters and a right hypotropia of 6 prism diopters in primary position. Her external appearance and ocular motility photographs were taken postdilation and are shown in Figure 16.1. Discrete superficial punctated keratitis was seen OU. Fundus examination was within normal limits.

PRO: LOW DOSE ORBITAL RADIATION THERAPY IS A USEFUL ALTERNATIVE IN THE TREATMENT OF THYROID EYE DISEASE

Reid Longmuir

External beam low dose orbital radiotherapy has been used in thyroid eye disease for many years. Radiotherapy has nonspecific antiinflammatory and presumed suppression effects on the offending lymphocytes infiltrating the orbit. At our institution, like most others we use low dose orbital treatment of 20 gray (Gy) fractionated over 2-weeks (10 day sessions). Although cataract, radiation retinopathy, and radiation optic neuropathy may occur fortunately these are rare complications.

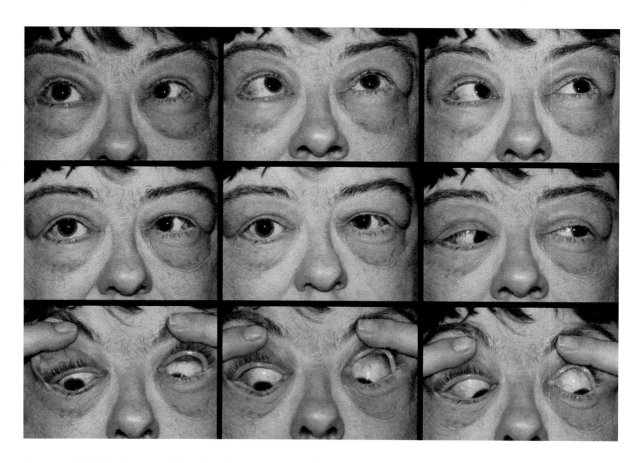

Figure 16.1 Motility photograph showing a large esotropia with bilateral abduction deficit, bilateral elevation deficit, and proptosis with marked periorbital edema.

Wei et al. (1) systematically evaluated the efficacy of orbital radiotherapy in the treatment of thyroid eye disease based upon a review of the Cochrane Library, Medline, Embase, and the Chinese Biomedical Database. A total of 18 studies were included in the meta-analysis (8 cohort and 10 randomized studies). These authors concluded that orbital radiotherapy alone was significantly more effective than control and as effective as oral corticosteroids. The combination of both orbital radiotherapy and oral corticosteroids was "markedly more effective" than other treatment modalities.

Bradley et al. (2) also reviewed the medical literature databases to identify all published reports relating to orbital radiation treatment for thyroid eye disease. This "technology assessment" for the American Academy of Ophthalmology (AAO)included 1) cases with original data, 2) if a case series or uncontrolled trial included at least 100 subjects, and 3) randomized clinical trials of any size. Abstracted data included study and patient characteristics, treatment response, and safety information. In this assessment there were 14 studies that included five observational studies and nine randomized controlled trials. The results showed three observational studies with overall favorable treatment outcomes from 40% to 97% of patients and three observational studies provided intermediate-term safety data. The risk of definite radiation retinopathy was 1–2% within 10 years after treatment and there was no increased risk of secondary malignancy or premature death. The nine randomized trials were qualitatively heterogeneous and unfortunately patients with thyroid optic neuropathy typically were excluded from these randomized trials and therefore no comment can be made for this indication. Three of the randomized trials were sham controlled and none of these showed that orbital radiation was more efficacious than sham irradiation for improving proptosis, lid fissure, or soft tissue changes. Two of three sham-controlled randomized trials demonstrated improved vertical range of motion. The systematic review concluded that "the effect of orbital radiation on Graves ophthalmopathy is limited by the lack of standardization and variable quality of published reports" but that "extraocular motility impairment may improve with radiotherapy, although the evidence of a treatment effect is mixed in clinical trials." The technology assessment also reported that "Future studies are needed to determine if a potentially beneficial motility effect results in improved patient function and quality of life. Level I evidence indicates that proptosis, eyelid retraction, and soft tissue changes do not improve with radiation treatment. The efficacy of orbital radiation for compressive optic neuropathy resulting from Graves ophthalmopathy has not been investigated in clinical trials and merits further study."

Given the studies performed and reviewed to date, it appears that low dose orbital radiation therapy is an option for compressive optic neuropathy especially in patients who are not good surgical or steroid candidates. Although the evidence is mixed for ophthalmoplegia, many of our patients self report subjective improvement in comfort and have objective improvement in ophthalmoplegia.

Because the current evidence is conflicting on the efficacy of radiotherapy future studies will be necessary to help answer the question. The Combined Immunosuppression and Radiotherapy in Thyroid Eye Disease (CIRTED) trial was designed to investigate the efficacy of radiotherapy and azathioprine in combination with a standard course of oral prednisolone in patients with active thyroid eye disease. Patients will be randomized to azathioprine or oral placebo and radiotherapy or sham-radiotherapy in this multicenter controlled clinical trial with the primary outcome measure being improvement in disease severity as assessed by a composite binary measure at 12 months. Secondary end-points include quality of life scores and health economic measures.

REFERENCES

1. Wei RL, Cheng JW, Cai JP. The use of orbital radiotherapy for Graves' ophthalmopathy: quantitative reviewof the evidence. Ophthalmologica 2008; 222(1): 27–31.
2. Bradley EA, Gower EW, Bradley DJ et al. Orbital radiation for graves ophthalmopathy: a report by the American Academy of Ophthalmology. Ophthalmology 2008; 115(2): 398–409.

BIBLIOGRAPHY

Bartalena L, Marcocci C, Manetti L et al. Orbital radiotherapy for Graves' ophthalmopathy. Thyroid 1998; 8: 439–44.

Kahaly G, Beyer J. Immunosuppressant therapy of thyroid eye disease. Klin Wochenschr 1988; 66: 1049–59.

Marcocci C, Bartalena L, Bogazzi F, Bruno-Bossio G, Pinchera A. Role of orbital radiotherapy in the treatment of Graves' ophthalmopathy. Exp Clin Endocrinol 1991; 97: 332–37.

Petersen IA, Kriss JP, McDougall IR, Donaldson SS. Prognostic factors in the radiotherapy if Graves' ophthalmopathy. Int J Radiat Oncol Biol Phys 1990; 19: 259–64.

Nakahara H, Noguchi S, Murakami N. Graves' ophthalmopathy: MR evaluation of 10-Gy vs. 24-Gy irradiation combined with systemic corticosteroids. Radiology 1995; 196: 857–62.

Kahaly GJ, Rösler H-P, Pitz S, Hommel G. Low- versus high-dose radiotherapy for Graves' ophthalmopathy: a randomized, single blind trial. J Clin Endocrinol Metab 2000; 85: 102–8.

Bartalena L, Marcocci C, Chiovato L et al. Orbital cobalt irradiation combined with systemic corticosteroids for Graves' ophthalmopathy: comparison with systemic corticosteroids alone. J Clin Endocrinol Metab 1983; 56: 1139–44.

Miller ML, Goldberg SH, Bullock JD. Radiation retinopathy after standard radiotherapy for thyroid-related ophthalmopathy. Am J Ophthalmol 1991; 112: 600–1.

Kinyoun JL, Kalina RE, Brower SA, Mills RP, Johnson RH. Radiation retinopathy after orbital irradiation for Graves' ophthalmopathy. Arch Ophthalmol 1984; 102: 1473–6.

Orcutt JC, Kinyoun JL. Radiation and Graves' ophthalmopathy (letter). J Clin Endocrinol Metab 1995; 80: 2543.

Nygaard B, Specht L. Transitory blindness after retrobulbar irradiation of Graves' ophthalmopathy. Lancet 1998; 351: 725–6.

Snijders-Keilholz A, De Keizer RJW, Goslings BM et al. Probable risk of tumor induction after retro-orbital irradiation for Graves' ophthalmopathy. Radiother Oncol 1996; 38: 69–71.

Blank LECM, Barendsen GW, Prummel MF et al. Probable risk of tumor induction after retro-orbital irradiation for Graves' ophthalmopathy (letter). Radiother Oncol 1996; 40: 187–8.

Gorman CA. Double-blind prospective controlled study of orbital radiotherapy for Graves' ophthalmopathy: results at one year. J Endocrinol Invest 1999; 22 6): 102.

Wiersinga WM, Prummel MF. Retrobulbar radiation in Graves' ophthalmopathy. J Clin Endocrinol Metab 1995; 80: 345–7.

Rajendram R, Lee RW, Potts MJ et al. Protocol for the combined immunosuppression & radiotherapy in thyroid eye disease (CIRTED) trial: a multi-centre, double-masked, factorial randomised controlled trial. Trials 2008; 31; 9: 6.

CON: RADIATION THERAPY IS NOT HELPFUL IN THE TREATMENT OF THYROID EYE DISEASE

Karl Golnik

Thyroid eye disease (TED) is an autoimmune inflammatory disorder whose underlying cause remains unknown. The clinical signs, however, are characteristic and may include conjunctival injection, chemosis, eyelid retraction, eyelid lag, proptosis, restrictive extraocular myopathy, and optic neuropathy. Fortunately most patients (95%) have mild or moderate TED and do not progress to the severe form which includes optic neuropathy and/or severe proptosis with corneal decompensation. The disease activity in the two eyes may be remarkably asymmetric. Although typically associated with hyperthyroidism, thyroid eye disease may accompany hypothyroidism or rarely Hashimoto thyroiditis; in about 10% of patients characteristic eye findings occur without objective evidence of thyroid dysfunction ("euthyroid Graves disease"). The course of the eye disease does not necessarily parallel the activity of the thyroid gland or the treatment of thyroid abnormalities.

Recent data suggests that tight thyroid control may be of benefit to orbital disease. However, in some studies, treatment with radioactive iodine (RAI) has been associated with an exacerbation of orbital disease, and some authorities suggest that concurrent corticosteroid therapy may reduce the incidence of this effect. Other authorities disagree and recommend

proceeding with RAI. Smoking cigarettes has been identified as a risk factor for the progression of TED. Thus, patients should be encouraged to quit smoking. Choice of therapy depends on the signs and symptoms present. Many patients require only supportive care for ocular symptoms, such as topical ocular lubricant ointment at night and artificial tears during the day. If there is significant chemosis and pain, corticosteroid therapy may be effective (1.0–1.5 mg/kg prednisone), but the side effects of chronic corticosteroid therapy (>2 months) typically outweigh the benefit. Taping the eyelids shut at night may also be effective in patients with lagophthalmos. For acute cases with severe corneal exposure tarsorrhaphy may be necessary. Recession of the upper and lower eyelid retractors may be done for chronic lid retraction. Eyelid surgery should be deferred if orbital surgery or eye muscle surgery is contemplated. The diplopia associated with TED is related to progressive muscular fibrosis. Although short-term corticosteroid therapy may help control active inflammation, no specific treatment can reverse fibrosis. In acute cases, double vision associated with restrictive strabismus can be eliminated by occlusion. After the deviation becomes stable, eye muscle surgery may achieve realignment. Optical realignment may be possible with spectacle prisms ground into the lenses or Fresnel press-on prisms. The presence of optic nerve dysfunction requires prompt therapeutic intervention. In most cases, a trial of moderately high doses of oral corticosteroids may result in substantial improvement in optic nerve function but this typically recurs when the steroids are stopped. The definitive treatment is to surgically decompress the optic nerve in the orbital apex. Although any of the four orbital walls may be decompressed, removal of the posterior medial wall is usually most effective. This surgical maneuver may be accomplished endoscopically, as an external ethmoidectomy (through the caruncle), or through the maxillary sinus (Caldwell-Luc). When proptosis is the major feature, removal of the orbital floor (by way of an eyelid or conjunctival incision or through the maxillary sinus) and possibly the lateral wall may help decrease the globe prominence. Patients need to be aware that decompression surgery may adversely affect ocular motility and eyelid position. Thus, eyelid and extraocular muscle surgery should be deferred if orbital decompressive surgery is contemplated.

Debate exists as to the efficacy of radiotherapy in TED. Theoretically, orbital radiation therapy may be efficacious because the activated T-cells and fibroblasts in TED are radiosensitive, and therapy can be delivered locally without the systemic side effects associated with corticosteroids. (AAO paper) Orbital radiotherapy for TED usually is administered as a 20-Gy cumulative dose delivered in 10 divided fractions over 2-weeks.

Only three trials have looked solely at radiation *versus* no treatment. All three compared radiation to sham radiation. Mourits and associates reported 59 patients with moderately severe TED, 30 of whom were randomized to bilateral orbital radiation and 29 of whom received bilateral sham orbital radiation. Eighty-two percent of patients treated with orbital radiation had improved motility compared with 27% of sham-irradiated subjects ($P = .004$). No differences were seen between treatment groups for change in lid fissure, soft tissue swelling, proptosis, or subjective eye score, at 24 weeks of follow-up. Although randomization produced groups that were similar in most ways, there was a difference in pretreatment motility; 12 of 30 orbital-radiation subjects had diplopia in all positions of gaze, compared with only 4 of 30 sham-radiated subjects. Gorman and associates randomized 42 subjects with bilateral mild-to-moderate TED to receive orbital radiation to one orbit and sham radiation to the other. Forty-five percent of study subjects had received systemic corticosteroids before orbital radiation. No clinically or statistically significant difference between the treated and untreated orbit was observed in volume of extraocular muscle and fat, proptosis, range of extraocular muscle motion, area of diplopia fields, and lid fissure width at 6 months. At 12 months, muscle volume and proptosis improved slightly more in the orbit that was treated first. No significant improvement was reported in a follow-up study of these same patients 3 years later. Prummel and associates reported 88 patients with mild TED who were randomized to bilateral orbital radiation (44 subjects) compared with bilateral sham irradiation (44 subjects).(1) Patients treated with irradiation gained 6.0° of globe depression (95% confidence interval, 2.0–10.1) and had improved ocular range of motion (mean difference, 370 mm2; 95% confidence interval, 1–739). There was no statistically significant difference between treatment groups for change in eye elevation, adduction, or abduction. Lid aperture, proptosis, and clinical activity score were not significantly different between groups. Despite the mild clinical improvements, the results of a quality of life questionnaire did not indicate any differences between the two groups.

A panel appointed by the American Academy of ophthalmology reported results of an extensive literature review (2) including only observational studies with at least 100 patients (five studies) and randomized controlled clinical trials (nine studies). They concluded that lid retraction, proptosis, and soft tissue changes do not respond to radiation and there is mixed results with extraocular motility. Patients with compressive optic neuropathy were generally excluded from trials and thus further studies in this group of patients are necessary. Interpretation of many reports is complicated by concomitant use of corticosteroids, no mention of whether corticosteroids were used, concomitant use of intravenous immunoglobulin, no data on severity of disease at onset of treatment, and no data on length of follow-up. Furthermore, two of the nine trials compared results of different protocols of radiation delivery as opposed to whether radiation was better than observation. Finally, one must consider potential complications of radiation. Tumors in the site of radiation are extremely rarely reported but radiation retinopathy has been reported in 1–2% of patients.

It would appear that radiation may help some measured parameters in patients with mild to moderate TED but overall it does not significantly improve their condition. Thus, I would not recommend orbital radiation for treatment in patients with mild or moderate TED. However, patients with severe TED with optic neuropathy and/or extremely active periorbital/orbital edema with massive proptosis and corneal exposure have not been systematically studied. I believe orbital radiation may benefit this group but usually reserve radiation for patients who either refuse or are considered a poor risk for surgical orbital decompression.

REFERENCES

1. Prummel MF, Terwee CB, Gerding MN et al. A randomized controlled trial of orbital radiotherapy versus sham irradiation in patients with mild Graves' ophthalmopathy. J Clin Endocrinol Metab 2004; 89: 15–20.
2. Bradley EA, Gower EW, Bradley DJ et al. Orbital irradiation for Graves ophthalmopathy: a report from the American Academy of Ophthalmology. Ophthalmology 2008; 115: 398–409.

BIBLIOGRAPHY

Bartalena L, Marcocci C, Bogazzi F et al. Relation between therapy for hyperthyroidism and the course of Graves' ophthalmopathy. N Engl J Med 1998; 338: 73–8.
Sisson JC, Schipper MJ, Nelson CC, Freitas JE, Frueh BR. Radioiodine therapy and Graves' ophthalmopathy. J Nucl Med 2008; 49(6): 923–30.
Prummel MF, Wiersinga WM. Smoking and risk of Graves' disease. JAMA 1993; 269: 479–82.
Mourits MP, van Kempen-Harteveld ML, Garcia MB et al. Radiotherapy for Graves' orbitopathy: randomised placebo controlled study. Lancet 2000; 355: 1505–9.
Gorman CA, Garrity JA, Fatourechi V et al. A prospective, randomized, double-blind, placebo-controlled study of orbital radiotherapy for Graves' ophthalmopathy. Ophthalmology 2001; 108: 1523–34.
Gorman CA, Garrity JA, Fatourechi V et al. The aftermath of orbital radiotherapy for Graves' ophthalmopathy. Ophthalmology 2002; 109: 2100–7.
Bartalena L, Marcocci C, Tanda ML et al. Cigarette smoking and treatment outcomes in Graves ophthalmopathy. Ann Intern Med 1998; 129: 632–5.
Marcocci C, Bartalena L, Rocchi R et al. Long-term safety of orbital radiotherapy for Graves' ophthalmopathy. J Clin Endocrinol Metab 2003; 88: 3561–6.
Wakelkamp IM, Tan H, Saeed P et al. Orbital irradiation for Graves' ophthalmopathy: is it safe? A long-term follow-up study. Ophthalmology 2004; 111: 1557–62.
Marquez SD, Lum BL, McDougall IR et al. Long-term results of irradiation for patients with progressive Graves' ophthalmopathy. Int J Radiat Oncol Biol Phys 2001; 51: 766–74.

Schaefer U, Hesselmann S, Micke O et al. A long-term follow-up study after retro-orbital irradiation for Graves' ophthalmopathy. Int J Radiat Oncol Biol Phys 2002; 52: 192–7.

Bartalena L, Marcocci C, Chiovato L et al. Orbital cobalt irradiation combined with systemic corticosteroids for Graves' ophthalmopathy: comparison with systemic corticosteroids alone. J Clin Endocrinol Metab 1983; 56: 1139–44.

Antonelli A, Saracino A, Alberti B et al. High-dose intravenous immunoglobulin treatment in Graves' ophthalmopathy. Acta Endocrinol (Copenh) 1992; 126: 13–23.

Gerling J, Kommerell G, Henne K et al. Retrobulbar irradiation for thyroid-associated orbitopathy: double-blind comparison between 2.4 and 16 Gy. Int J Radiat Oncol Biol Phys 2003; 55: 182–9.

Kahaly GJ, Rosler HP, Pitz S, Hommel G. Low- versus high-dose radiotherapy for Graves' ophthalmopathy: a randomized, single blind trial. J Clin Endocrinol Metab 2000; 85: 102–8.

Ng CM, Yuen HK, Choi KL et al. Combined orbital irradiation and systemic steroids compared with systemic steroids alone in the management of moderate-to-severe Graves' ophthalmopathy: a preliminary study. Hong Kong Med J 2005; 11: 322–30.

Prummel MF, Mourits MP, Blank L et al. Randomized double-blind trial of prednisone versus radiotherapy in Graves' ophthalmopathy. Lancet 1993; 342: 949–54.

SUMMARY

There have been several prospective, masked clinical trials with sham controls for orbital radiotherapy in thyroid eye disease. Unfortunately the evidence is conflicting and remains controversial and the quality of the reporting is variable. Part of the problem with the evidence base is that the inclusion and exclusion criteria for these studies bias the results. Patients with active inflammatory disease (e.g., red, hot, swollen eyes) or compressive optic neuropathy would theoretically be the patients who would most benefit from radiotherapy. These patients however were purposefully not studied in many of the published reports. Thus, the controversy remains in our opinion unresolved on the utility of radiation therapy in thyroid eye disease. The data would suggest that patients with chronic or mild disease or "quiet" eyes in the fibrosis phase of the disorder are unlikely to benefit from orbital radiation therapy. There probably is a subset of patients however who still might benefit from consideration for orbital radiotherapy and an appropriate risk to benefit and informed consent discussion should take place for these patients.

17 Should I do topical pharmacologic testing in the Horner syndrome?

A 50-year-old male with no prior medical history, but with a history of heavy smoking, presents to the local ophthalmology office on referral from his internist for large right pupil. The patient has not noticed it previously. He has no other complaints. Visual acuity is 20/20 in each eye, confrontation visual fields, motility, are normal. The pupils are shown below in Figures 17.1 and 17.2, with anisocoria greater in dark than in light. There appears to be a subtle, but uncertain, dilation lag on the left. There is no prior history of surgery or trauma to suggest this anisocoria is due to iris sphincter damage. The patient wants to know if this could be serious and how soon he can find out.

Figure 17.1 Pupils under light condition. The pupils are nearly equal.

Figure 17.2 Pupils under dark condition demonstrating that the left pupil does not dilate as well as the right pupil, resulting in about 1.5 mm of anisocoria in the dark.

PRO: PHARMACOLOGIC TESTING IS USEFUL IN THE EVALUATION OF POSSIBLE HORNER'S SYNDROME

Fiona Costello

Horner's Syndrome (HS) refers to a disturbance of oculosympathetic innervation, which is characterized by the classic triad of miosis, ptosis, and forehead anhydrosis. The patient in this case demonstrates many of the cardinal features of HS including anisocoria, which is greatest in darkness. Not uncommonly, as in this case, the eye with the larger pupil is initially thought to be the abnormal one, when in fact the problem pertains to the pupil that fails to dilate properly. Because patients are often examined in bright lighting conditions, this subtle clinical sign may be overlooked, particularly if it occurs in isolation. Other clinical signs of HS include lower lid elevation or "upside down ptosis", apparent enophthalmos, increased accommodation, and ocular hypotony. This condition can be congenital or acquired, and patients may manifest one or more of the aforementioned features. In congenital HS, ipsilateral heterochromia of the iris can be seen. The diagnosis of HS is important not to miss, because it can be the harbinger of potentially life threatening disorders including arterial dissection, neuroblastoma, and lung malignancies.

Understanding the anatomy of the sympathetic nervous system is germane to identifying the cause of HS. The pupil dilator muscle in controlled by sympathetic innervation, which proceeds from the ipsilateral hypothalamus, through the lateral tegmentum of the brainstem, and into the intermediolateral gray matter of the spinal cord at the ciliospinal center of Budge (C8–T2 segments). The sympathetic fibers then course to the superior cervical ganglion, the carotid plexus, and the ophthalmic branch of the trigeminal nerve, before finally reaching the pupil through the long ciliary nerves.(1) The sympathetic chain is vulnerable to injury along many points involving the first, second, and third-order neurons. The first order neuron, which begins in the posterolateral hypothalamus, may be damaged by central lesions such as primary brain tumors, demyelinating plaques, and strokes (lateral medullary infarct). The second order neuron is in the intermediolateral cell column of the low cervical and upper thoracic cord; and may be injured by lesions of the neck, including Pancoast tumor, and birth trauma causing a brachial plexus palsy. Damage to the third order neuron, which arises from the superior cervical ganglion, can result from internal carotid artery dissection, cavernous sinus fistula, and cluster headaches.

The diagnosis of HS can be confirmed by instilling dilute (4–10%) cocaine in each eye and comparing the extent of anisocoria 45 minutes later. Cocaine blocks the presynaptic uptake of norepinephrine at the neuromuscular junction in the pupil dilator muscle. A normal pupil will dilate fully whereas the affected pupil will fail to dilate completely. (1) Kardon and colleagues (3) evaluated the effectiveness of cocaine testing in 119 patients with HS and 50 normal subjects. The chances of having Horner's syndrome increased with the amount of cocaine induced anisocoria, such that a postcocaine anisocoria value of 0.8 mm gave a mean odds ratio of approximately 1,050:1 that Horner's syndrome was present (lower 95% confidence limit = 37:1). Cocaine testing is not infallible, however, as shown by Van der Wiel and colleagues (4), who studied 20 patients with suspected HS and 20 controls. They noted that a difference in mydriatic response to cocaine of > 1.0 mm between the two eyes occurred only in patients with HS; whereas a mydriatic response < 1.0 mm correlated to only a 60% chance that the patient did not have HS.

In their study, the investigators noted no relationship between the magnitude of the response to cocaine and the site of the lesion in the sympathetic system.(4) Apraclonidine (0.5–1%) has also been used to confirm the diagnosis of HS.(2, 5) This ocular hypotensive agent acts as an alpha-1 receptor agonist, with little or no effect on a normal pupil. Because patients with HS generally have denervation super-sensitivity of the affected pupil, instillation of apraclonidine solution will cause a reversal of anisocoria. However, this pharmacological test has also been associated with false-negative results, and is not 100% sensitive to the diagnosis of HS.

Hydroxyamphetamine (1%) is used to differentiate preganglionic from postganglionic causes of HS. This solution releases norepinephrine into the synaptic cleft from intact presynaptic postganglionic nerve terminals. Hydroxyamphetamine instilled into an eye with Horner syndrome with intact postganglionic fibers (first- or second-order neuron lesions) dilates the pupil to an equal or greater extent than the normal pupil. An eye with HS due to damaged postganglionic fibers (third-order neuron lesions) does not dilate as well as the normal pupil after hydroxyamphetamine. No pharmacological agent will differentiate a first-order neuron from a second order neuron Horner's syndrome.

While pharmacological testing does help localize the potential site of an oculosympathetic pathway lesion, current agents are not 100% sensitive. For this reason, patients with equivocal results should undergo appropriate investigations, because of the potential risk to miss a dire diagnosis. Pharmacological testing however does have merit in the evaluation of HS for a number of reasons. First, physiological anisocoria can mimic Horner's and occurs in approximately 20% of people. A negative cocaine or apraclonidine test may not obviate the need for neuroimaging or chest imaging for all patients, but may provide some reassurance for patients with intermittent or longstanding anisocoria. It would be a costly endeavor to image all patients who manifest any degree of anisocoria greater in darkness, due to the remote chance that they might harbor a lesion of the sympathetic pathway. Secondly, neuroimaging is not 100% sensitive to the detection of all lesions in the sympathetic pathway, particularly if the imaging study is not directed at the appropriate anatomical region or if the wrong imaging protocol is employed. The sympathetic nervous system covers a fairly large area of "anatomical real-estate", and to fully account for all possible first, second, and third order neuron lesions would require gadolinium enhanced head and neck magnetic resonance imaging (MRI); head and neck MR angiography (MRA) [or computed tomography angiography (CTA)]; and MRI or CT imaging of the thorax. To implement all protocols at all times for all patients with suspected HS would be quite time consuming, not to mention cost-ineffective. Digre and colleagues performed MR imaging in 33 patients with HS, including 13 preganglionic and 20 postganglionic cases as determined by pharmacological testing. Imaging abnormalities were noted in half the patients with preganglionic HS;

and three of 20 patients with postganglionic HS. Routine scanning of patients with postganglionic HS with cluster headaches was not helpful in this small series.

In the case example provided, one would worry about a lung lesion such as a Pancoast tumor in light of the patient's age and smoking history. It would therefore be reasonable to take a more focused approach in the evaluation of this patient, in lieu of implementing an "everything but the kitchen sink" series of imaging studies. Cocaine or apraclonidine could be used to confirm the suspected diagnosis of HS, if there was any doubt about the diagnosis. However, if the clinical examination was unequivocal, hydroxyamphetamine could be used to distinguish whether the sympathetic lesion is likely pre or postganglionic. Evidence supporting a preganglionic lesion would prompt one to examine the supraclavicular region for palpable nodes; and proceed with a chest X–ray. In this case, a positive test result with the most simple and easily acquired imaging study could prevent unnecessary imaging studies of the entire sympathetic chain. In the event that the preliminary chest imaging is negative, one could then proceed with more detailed vascular and cranial imaging with MRI; and consider a more detailed chest imaging study with CT or MRI scanning.

A third reason to consider pharmacological testing as an adjunct to the evaluation of HS is that not all cases of HS are due to a structural lesion. Other causes for HS include brachial plexus palsy, dysautonomia syndromes, lumbar epidural anesthesia, and cluster headaches, to name a few. In these cases, imaging studies will not provide insights regarding cause; whereas pharmacological testing may help define potential mechanisms, unrelated to structural entities, which may explain the HS.

Thus, pharmacological testing can complement the evaluation of HS, by helping the clinician to decide where the culprit lesion is most likely to be found, and what imaging modality should be used to detect the cause of HS. The evaluation of HS requires understanding of the many potential mechanisms that can impact the oculosympathetic pathway; because, even in the modern imaging era, HS remains first and foremost a clinical diagnosis.

REFERENCES

1. Brazis PW, Masdeu JC, Biller J. Localization in Clinical Neurology, 3rd ed., Lippincott, Williams & Wilkins, 1996.
2. Bardorf C. Horner syndrome. eMedicine. http://www.emedicine.com/OPH/topic336.htm
3. Kardon RH, Denison DE, Borwn CK, Thompson HS. Critical evaluation of the cocaine test in the diagnosis of Horner's syndrome. Arch Ophthlamol 1990; 108: 1667–8.
4. Van der Wiel HL, Van Gijn J. The diagnosis of Horner's syndrome. Use and limitations of the cocaine test. J Neurol Sci 1986; 73: 311–6.
5. Mirzai H, Baser EF. Congenital Horner's syndrome and the usefulness of the apraclonidine test in its diagnosis. Indian I Ophthalmol 2006; 54: 197–9.

BIBLIOGRAPHY

Digre KB, Smoker WR, Johnston P et al. Selective MR imaging approach for evaluation of patients with Horner's syndrome. AJNR Am J Neuroradiol 1992; 13: 223–7.

Kawasaki A, Borruat FX. False negative apraclonidine test in two patients with Horner syndrome. Klin Monatsbl Augenheilkd 2008; 225: 520–2.

Cremer SA, Thompson HS, Digre KB, Kardon RH. Hydroxyampetamine mydriasis in Horner's syndrome. Am J Ophthalmol 1990; 110: 71–6.

Shin RK, Galetta SL, Ting TY, Armstrong K, Bird SJ. Ross syndrome plus: beyond horner, Holmes–Adie, and harlequin. Neurology 2000; 55: 1841–6.

Rohrer JD, Schapira AH. Transient Horner's syndrome during lumbar epidural anaesthesia. Eur J Neurol 2008; 15: 530–1.

CON: PHARMACOLOGIC TESTING IS NOT NECESSARY IN THE EVALUATION OF POSSIBLE HORNER'S SYNDROME AND ONE SHOULD PROCEED DIRECTLY TO NEUROIMAGING

Nicholas Volpe

Horner's syndrome is a high stakes diagnosis. The subset of patients that have Horner's syndrome have potentially life threatening conditions including carotid dissection, neuroblastoma, lung tumors, and brainstem strokes. The clinician cannot afford to miss a diagnosis of Horner's syndrome, nor orchestrate a misdirected work up. If there is any suspicion for possibility of Horner's syndrome, that diagnosis should be pursued regardless of the results of pharmacologic testing. There are several different scenarios in which Horner's syndrome develops. Admittedly the vast majority of patients with isolated painless Horner's syndrome, in the adult population that have no other relevant history or symptoms, are going to turn out to have simple vasculopathic palsies that require no work up. There is clearly a subset of patients that have significant disease as a cause for their Horner's syndrome, including neuroblastoma in a child, carotid dissection, and lung tumor or brainstem stroke. The neuroophthalmologist plays a vital role in the recognition of this important clinical finding and steering the patient towards workup.

In the end, most Horner's syndromes are recognized by a classic clinical examination, which included ptosis and miosis. In the case presented there is anisocoria worse in the dark making the diagnosis of a Horner's syndrome more likely. While there is minimal apparent ptosis seen in the photographs, the patient maybe elevating the lid with his brow. Miosis generally takes on typical features that include worsening in the dark and/or dilation lag and the ptosis is often associated with lower eyelid ptosis which affectively raises the lower eyelid, narrows the palpebral fissure and gives the appearance of enophthalmos. This makes the clinical recognition of Horner's syndrome rarely difficult and most situations pharmacologic testing only turns out to be superfluous

and may very well lead the clinician to an erroneous conclusion. If there is even a 10% chance of a false negative or false positive result and the mistakes include missing diagnoses such as life threatening tumors and carotid dissections, the clinician simply cannot fail to act on his/her clinical suspicion that a Horner's syndrome is present and recommend a workup regardless of the results of pharmacologic testing. If you suspect it clinically and based on exam, then the diagnostic workup should be performed based on the patient's age and other symptoms as indicated. This workup may very well include imaging studies that span from the mediastinum of the chest to the brain and with attention to both vascular and soft tissue structures as well as the need to rule out lung tumors.

There are a number of reasons that pharmacologic testing can be associated with both false positives and false negatives. False negatives, of course, are the group of patients that we would be most concerned about. The clinical suspicion is there, you do the test, and it does not suggest a Horner's syndrome. What are the possible explanations? The first of course is that these tests are based on the development of denervation, suprasensitivity. You may be seeing the patients too soon in their course for denervation suprasensitivity to be demonstrated. There are as well many patients in whom a normal pupil will not dilate well during a cocaine test (particularly dark irides), which may confuse the results of the test. The examiner must identify small degrees of postcocaine anisocoria which can be difficult. Corneal epithelial disease may cause asymmetric penetration of the medications into the anterior segment and affect the results of the pharmacologic testing. Clinicians may also attempt to localize with pharmacologic testing (hydroxyamphetamine) in order to subgroup the Horner's as either pre or postganglionic with the presumption that preganglionic lesions are more concerning (brainstem, lung localization). Again a false positive or negative result here could cause a misdirected or inadequate work up.

A clinician puts himself or herself at risk to depend on pharmacologic testing which has an unacceptably high rate of false negatives and positives. This is not to say that every patient with a Horner's syndrome needs an extensive workup, but the clinician should come to a conclusion based on his or her impression of the history, the clinical exam of the eyelids and pupils and pursue a workup accordingly and not potentially get bogged down in the inaccuracies of pharmacologic testing.

BIBLIOGRAPHY

Maloney WF, Younge BR, Moyer NJ. Evaluation of the causes and accuracy of pharmacologic localization in Horner's syndrome. Am J Ophthalmol 1980; 90(3): 394–402.

Newman NM, Levin PS. Testing the pupil in Horner's syndrome. Arch Neurol 1987; 44(5): 471.

Thompson HS. Diagnosing Horner's syndrome. Trans Sect Ophthalmol Am Acad Ophthalmol Otolaryngol 1977; 83(5): 840–2.

Thompson HS. Pharmacologic localization in Horner's syndrome. Am J Ophthalmol 1981; 91(3): 416–7.

Van der Wiel HL, Van Gijn J. The diagnosis of Horner's syndrome. Use and limitations of the cocaine test. J Neurol Sci 1986; 73(3): 311–6.

van der Wiel HL, van Gijn J. The diagnosis of Horner's syndrome. Clin Neurol Neurosurg 1988; 90(2): 103–8.

Van der Wiel HL, Van Gijn J. Localization of Horner's syndrome. Use and limitations of the hydroxyamphetamine test. J Neurol Sci 1983; 59(2): 229–35.

SUMMARY

In the past, topical cocaine (for confirmation of the Horner syndrome) and topical hydroxyamphetamine (for localization to the pre or postganglionic ocular sympathetic neuron) has been recommended in patients with anisocoria and the suspected diagnosis of the Horner syndrome. Pharmacologic confirmation and topographic localization helps localize and direct the neuroimaging and tells the radiologist where to look carefully for a lesion. Unfortunately, the pharmacologic testing is not 100% sensitive or specific and the possible etiologies for a Horner syndrome (e.g., neuroblastoma in a child and carotid dissection or Pancoast tumor in an adult) are potentially life threatening. Imaging the entire sympathetic axis (e.g., MRI from hypothalamus to the C8–T2 level in the chest) is expensive and time consuming. Topical cocaine and topical hydroxyamphetamine are becoming more difficult if not impossible to find and maintain in the clinic. Topical apraclonidine may be a reasonable alternative to pharmacologic testing with cocaine although it does not localize the lesion to preganglionic or postganglionic location. Apraclonidine is also an easier test to interpret as the smaller Horner syndrome pupil generally becomes larger and the normal larger pupil becomes smaller resulting in a reversal of the anisocoria. The ptosis also often reverses. We recommend pharmacologic confirmation of the Horner syndrome with apraclonidine followed by imaging of the entire ocular sympathetic axis for patients with either apraclonidine confirmed Horner syndrome or high clinical suspicion for the diagnosis. Patients with physiologic anisocoria and negative apraclonidine testing can be observed but we recommend documentation in the record of the findings. Patients with a high clinical suspicion for Horner syndrome in the acute phase(e.g., dilation lag) or who have equivocal pharmacologic testing probably should be considered for neuroimaging as the stakes are high for missing a carotid dissection or other structural lesion in the Horner syndrome.

18 Should a patient with giant cell arteritis have a fluorescein angiogram?

An 81-year-old woman with unremarkable past medical history presents with acute decreased vision OD beginning 2 days ago. She has no headache, no scalp tenderness, and no jaw claudication but complain of loss of appetite for the last 2 months without any weight loss. Her erythrocyte sedimentation rate is 45 mm/hr and her CRP is 0.5 mg/dl (normal < 0.5). Her vision is count fingers OD and 20/25 OS. Goldman perimetry revealed a dense inferonasal altitudinal defect (Figure 18.1). Optic disc edema was present OD (Figure 18.2) but fundus examination was completely normal OS. There is a marked choroidal perfusion deficit in the medial posterior choroidal distribution OD shown in Figure 18.3.

PRO: A FLUORESCEIN ANGIOGRAM IS USEFUL IN THE EVALUATION OF SUSPECTED GIANT CELL ARTERITIS

Fiona Costello

This case highlights the importance of distinguishing whether AION is of the artertic *versus* nonarteritic type for any given patient. Arteritic AION is the most common cause of blindness in patients with giant cell arteritis (GCA), and recognition of this clinical syndrome is critical to prevent subsequent blindness in the fellow eye and avoid other nonophthalmic complications of GCA. Ultimately, the treating physician must rely on details of the history and clinical examination to determine whether empirical

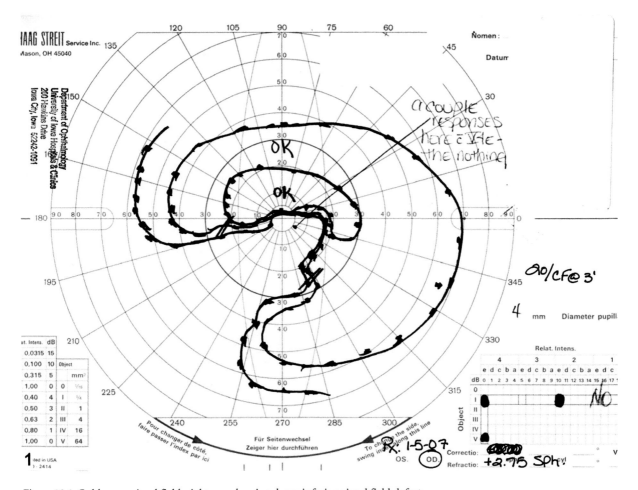

Figure 18.1 Goldmann visual field, right eye, showing dense inferior visual field defect.

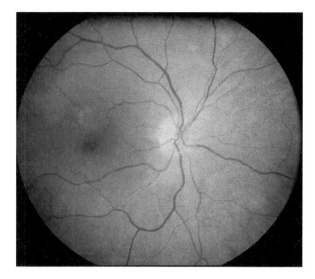

Figure 18.2 Optic nerve photography, right eye, showing pallid optic disc edema.

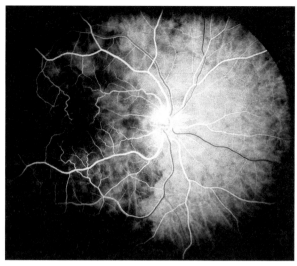

Figure 18.3 Fluorescein angiography, right eye, showing an area of choroidal nonperfusion temporally.

treatment is needed with high dose corticosteroid therapy and to identify which patients need to undergo a temporal artery biopsy.

In the case example provided, there are several features of the examination that serve as potential "red flags" for the clinician.

Age: First and foremost is the patient's advanced age of 81 years. The risk of arteritic AION increases with age which is worrisome in this case.

Acute phase reactants: Secondly, in this case, the patient has equivocal acute phase reactant results. According to Miller's formula, the patient's serum erythrocyte sedimentation rate (ESR) should be equal to or less than $(81 + 10) /2 = 46$ mm per hour; whereas Hayreh's formula indicates that the upper limit of normal for this patient's ESR should be $22.1 + (0.81 \times 81) = 37$ mm per hour. Three Studies have shown that serum ESR has a sensitivity of only $76 - 86\%$ in detecting GCA, and that serum CRP is a more sensitive indicator for the diagnosis. Because the patient's serum CRP is not normal (< 0.5 mg/dl), and given that the serum ESR is also borderline in this case, I would be inclined to err on the side of caution and treat the patient as a GCA suspect.

Systemic symptoms: This patient did not present with a prior history of transient vision loss, headache, jaw claudication, scalp tenderness or diplopia to suggest the diagnosis or GCA, but she did report loss of appetite without associated weight loss. GCA remains a tenable diagnosis, even in the absence of systemic manifestations. In a study of 85 biopsy – proven cases of GCA, Hayreh demonstrated that $> 20\%$ of GCA patients with ocular manifestations have no systemic symptoms at all. Furthermore, serum ESR and CRP levels are often lower in patients with occult GCA than those with preceding systemic manifestations. Therefore, I would remain suspicious for the diagnosis of GCA in this patient and treat accordingly, while awaiting the results of the temporal artery biopsy.

Fundus findings: The fact that the patient demonstrates pallid optic disc edema is also concerning for possible GCA. Nonarteritic AION is generally typified by optic disc edema, a small physiological cup, and associated flame hemorrhages. Further to this point, the area of choroidal nonperfusion demonstrated temporally with fluorescein angiography (FA) in the right eye, would also compel me to arrange an urgent temporal artery biopsy. Studies have shown that FA can help differentiate arteritic from nonarteritic AION by showing choroidal filling defects, which may be patchy or generalized, and peripapillary or peripheral. Massive choroidal nonperfusion is highly indicative of arteritic AION. Siatkowski et al.(2)retrospectively studied angiograms in 19 patients with nonarteritic AION and 16 patients with arteritic AION. Patients with GCA had delayed dye appearance and abnormal choroidal filling times, with a mean dye appearance of 20.3 seconds as compared to 11.29 seconds in nonarteritic patients. The mean choroidal filling time was 29.7 seconds in GCA patients, and 12.9 seconds in patients without GCA. When 18 seconds was used as the cut-off time for choroidal filling, the sensitivity of FA for the diagnosis of GCA was 93% and the specificity was 94%. (3, 4) Hayreh et al.(1) showed that when FA was performed within a few days after the onset of AION, and showed choroidal filling defects in the regions supplied by the posterior ciliary arteries, it was highly suggestive of arteritc AION. In this study of 170 patients with biopsy—proven GCA, FA was also useful in disclosing cilioretinal artery occlusion in 21.8% patients, and central retinal artery occlusion in 14.1% patients with ocular manifestations of GCA. Therefore, the studies to date indicate that in patients with acute AION, the finding of delayed choroidal filling on FA should raise the index of suspicion for GCA.

In my opinion, patients with AION who have borderline acute phase reactants should undergo FA to determine the role of temporal artery biopsy for GCA. Fluorescein angiography performed in the acute setting can disclose useful information about choroidal filling patterns, and concomitant vessel involvement, which may be highly suggestive of GCA. In patients with clear systemic manifestations of GCA or in those with marked elevation in their serum ESR and CRP levels, FA may not be necessary; as the results will not likely dissuade the course of management, which should involve high – dose corticosteroid therapy and temporal artery biopsy. In patients, however, with equivocal test results FA can prove extremely beneficial in identifying GCA as an important ophthalmic diagnosis with dire consequences if left untreated.

REFERENCES

1. Hayreh SS, Podhajsky PA, Zimmerman B. Occult Giant Cell Arteritis: ocular manifestations. Am J Ophthalmol 1998; 125: 521–6.
2. Siatkowski RM, Gass JDM, Glaser JS et al. Fluoroscein angiography in the diagnosis of giant cell arteritis. Am J Ophthalmol 1993; 115: 57–63.
3. Miller NR, Newman NJ. Clinical Neuro-Ophthalmology. Lippincott, Williams & Wilkins, 2005: 2362–3.
4. Rahman W, Rahman FZ. Giant cell (temporal) arteritis: an overview and update. Surv Ophthalmol 2005; 50: 415–28.

BIBLIOGRAPHY

Levin LA, Arnold AC. Neuro-Ophthalmology: The Practical Guide. NewYork: Thieme, 2005: 187–97.
Miller A, Green M, Robinson D. Simple rule for calculating normal erythrocyte sedimentation rate. Br Med J (Clin Res Ed) 1983; 286: 266.
Hayreh SS, Podhajsky PA, Raman R, Zimmerman B. Giant cell arteritis: validity and reliability of various diagnostic criteria. Am J Ophthalmol 1997; 123: 285–96.
Parikh M, Miller N, Lee AG et al. Prevalence of a normal c-reactive protein with an elevated erythrocyte sedimentation rate in biopsy—proven giant cell arteritis. Ophthalmology 2006; 113: 1842–5.
Hayreh SS, Podhajsky PA, Zimmerman B. Ocular manifestations of giant cell arteritis. Am J Ophthalmol 1998; 125: 509–20.

CON: FLUORESCEIN ANGIOGRAPHY IS USUALLY NOT NECESSARY IN THE EVALUATION OF POSSIBLE GIANT CELL ARTERITIS

Eric Eggenberger

Giant cell arteritis is often challenging to diagnose. In classic cases, the review of systems indicates symptoms of a diffuse systemic disease, lab evaluation demonstrates elevation of ESR and CRP, and a temporal artery biopsy reveals an inflammatory infiltrate within the vessel wall with the presence of giant cells; however, cases often lack one or more of these classic features. Accordingly, ancillary testing has been used to assist in securing the diagnosis or increasing diagnostic sensitivity of GCA. In addition to superficial temporal artery ultrasound, position emission tomography imaging, and novel laboratory marker,

fluorescein angiography (FA) has been proposed as a helpful ancillary test with potential use in the diagnosis of GCA.

Not all patients with GCA develop visual symptoms, and in the group of visually asymptomatic GCA patients, FA would be low yield. In a cohort of 161 patients diagnosed with biopsy-proven GCA, visual ischemic complications were present in only 42 (26.1%). FA in cases of GCA with ophthalmological involvement may demonstrate choroidal or retinal vascular defects, which in addition to optic nerve head edema indicate a more widespread vasculopathy than simple nonarteritic anterior ischemic optic neuropathy. The presence of retinal findings such as cotton wool spots likewise indicates ischemia involving not only posterior ciliary arteries supplying the optic nerve head, but also retinal artery disease. It should be noted that several of these retinal features are readily apparent via ophthalmoscopy, and do not require FA for detection. Although such FA-related findings are important, they are not diagnostic (nor specific), and FA is associated with some degree of risk (ranging from transient nausea and dizziness, to allergic reactions producing itching and skin rash, or rare potentially fatal anaphylactiod reactions). The exact sensitivity of FA in the diagnosis of GCA remains unknown. In addition, these FA-related features resolve within days, limiting their practical usefulness in the clinic. Temporal artery biopsy also has a small risk of complications, but remains the gold standard diagnostic test.

Because FA is not risk-free, specific, sensitive or diagnostic of GCA, and given the existence of useful and sensitive noninvasive markers of GCA, we do not routinely employ FA in the diagnostic evaluation of GCA.

BIBLIOGRAPHY

Hayreh SS. Anterior ischaemic optic neuropathy: differentiation of arteritic from non-arteritic type and its management. Eye 1990; 4: 25–41.
Fineschi V. Fatal anaphylactic shock during a fluorescein angiography. Forensic Sci Int 1999; 1009: 137–42.
González-Gay MA, García-Porrúa C, Llorca J et al. Visual manifestations of giant cell arteritis. Trends and clinical spectrum in 161 patients. Medicine 2000; 79(5): 283–92.

SUMMARY

Most patients with giant cell arteritis (GCA) present with symptoms (e.g., headache, scalp tenderness, polymyalgia rheumatica, signs (e.g., pallid edema), and laboratory evidence (e.g., elevated erythrocyte sedimentation rate or C-reactive protein) for the diagnosis. Some patients with GCA however (e.g., posterior ischemic optic neuropathy) may benefit from a fluorescein angiogram. The finding of a choroidal filling defect in the setting of presumed posterior ischemic optic neuropathy (PION) is strongly suggestive of the diagnosis of GCA. There are risks (including anaphylactic shock and death) from fluorescein angiography and it is probably neither appropriate nor necessary to perform the test in every patient with suspected GCA. The clinician should consider fluorescein angiography as a possible adjunctive test in the evaluation for GCA but it cannot replace the "gold standard" diagnostic test, the temporal artery biopsy.

19   Does pseudotumor cerebri without papilledema exist?

A 30-year-old female presents to the local ophthalmologist complaining of increasing headache. She feels that her most recent eye care provider has given her the wrong prescription, and that the new glasses were the source of the headache. She now rates the headache at 7/10 on the pain scale. She acknowledged having gained 15 pounds in the previous 6 months, and also reported hearing a "whooshing" sound in her ears at night which corresponded to her pulse. She has no diplopia, and no history of transient vision loss or obscuration. On examination, the patient is 5'5", weighing 200 lbs. She has full motility. Her pupil examination is unremarkable, and her slit lamp examination is normal. Fundus examination reveals the optic nerves shown below (Figure 19.1 and 19.2). Her OCT of the RNFL (Figure 19.3) and Goldmann visual fields (Figures 19.4 and 19.5) are also included.

PRO: IDIOPATHIC INTRACRANIAL HYPERTENSION (IIH) WITHOUT PAPILLEDEMA DOES EXIST

Timothy J McCulley and Thomas N Hwang

To start, let's review the evolution of papilledema, which is defined as optic disc changes secondary to elevated intracranial pressure (ICP). Hyperemia due to capillary dilation is generally felt to be the first noticeable finding, followed by blurring of the peripapillary retinal nerve fiber layer (NFL). However, Hayreh

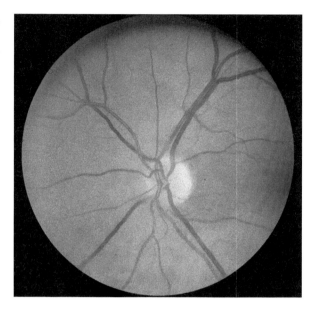

Figure 19.2 Optic nerve, left eye. Similar appearance to the right eye.

and Hayreh, using stereoscopic photography and angiography to assess experimentally-induced papilledema in nonhuman primates, described vascular changes occurring only after mild swelling of the nerve fiber layer.(1) Whichever appears first, slight blurring of the disk margin or hyperemia are both seen with early papilledema. These findings are subtle and may not easily be distinguished from normal. Also, the loss of spontaneous venous pulsations (SVPs) has been described as the earliest sign of papilledema. However, SVPs are normally not present in 20% or more of individuals and have also been shown to disappear in patients with disk edema due to causes other than papilledema.(2) The later more obvious findings include swelling/elevation of the optic disk, peripapillary hemorrhaging, and dilation of the retinal veins.

The subtlety of early papilledema makes it a challenge to determine with certainty whether papilledema is ever absent in patients with elevated ICP. In the case presented above, if high ICP was measured by lumbar puncture, one would have difficulty deciding whether the optic nerves are truly normal or have early signs of papilledema. In one sense, with known elevated ICP, the presence of papilledema has no diagnostic value, and one could argue that the mild hyperemia and blurred nasal margins in this case lack clinical relevance. Whether mild or absent papilledema truly preclude visual symptoms in patients with IIH will be addressed below.

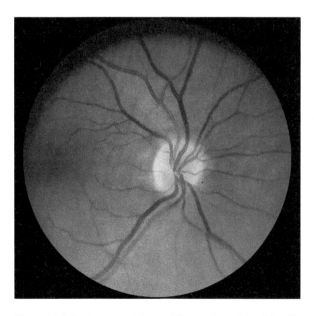

Figure 19.1 Optic nerve, right eye. The nasal portion of the disc is difficult to judge.

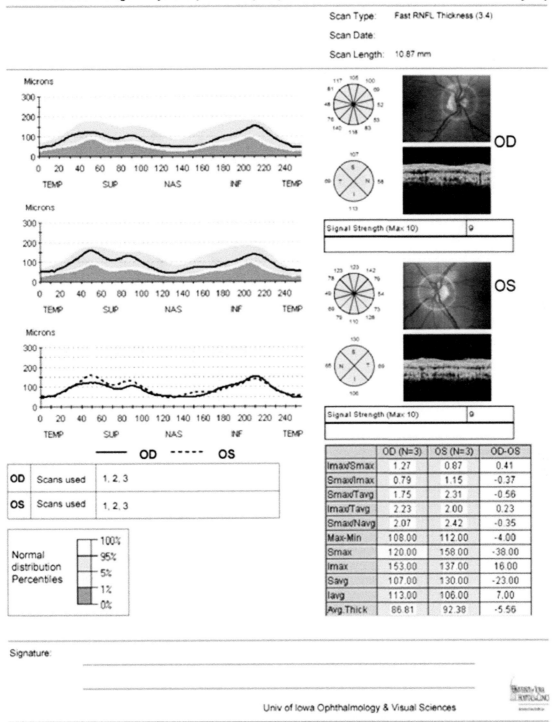

Figure 19.3 OCT of the RNFL, showing neither abnormal thickening, nor any thinning.

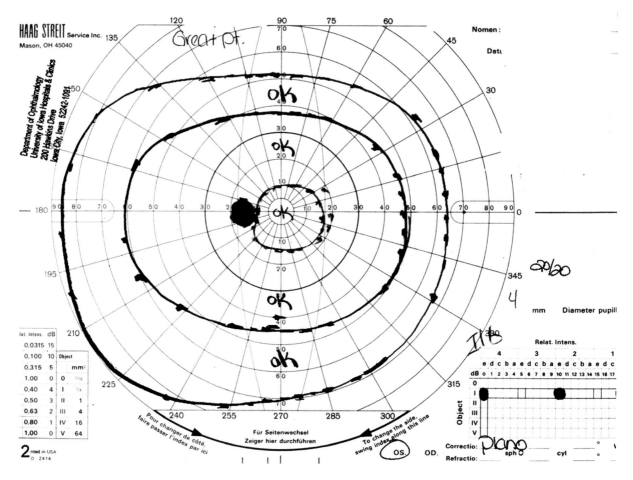

Figure 19.4 Goldmann visual field left eye, demonstrating normal field.

The specific question raised by this case is "Does idiopathic intracranial hypertension without papilledema (IIHWOP) exist?" This topic can be subdivided as follows:

1) What is the lag between the development of papilledema and ICP elevation?
2) With abnormally elevated ICP, are there patients who will never develop papilledema?
3) Do signs (optic atrophy) or symptoms (transient visual obscurations) ever develop in patients with elevated ICP in the absence of papilledema?

    Unfortunately, these are age-old questions addressed with little more than anecdotal case descriptions.

Papilledema does not develop instantaneously, and the lag between ICP elevation and appreciable papilledema will depend on the magnitude and rapidity of the ICP elevation. Cases with rapid elevations in ICP, such as following an intracranial hemorrhage, tend to develop papilledema more quickly.

Several investigators have described patients who developed papilledema within hours of ICP elevation. In 1969 Pagani described three patients with markedly elevated ICP following intracranial hemorrhages. All three developed papilledema within 2–4 hours of hemorrhage.(3) Similarly, Glowacki described two patients who developed papilledema within 5–8 hours of intracranial hemorrhage.(4) These examples illustrate that, even in extreme cases, there is some time lapse between ICP elevation and the development of papilledema. In conditions with less extreme elevation in ICP such as IIH, the time delay would be expected to be even longer so that during the earliest phase of the disease, patients with IIH will not have papilledema.

The question as to whether there are individuals who will never develop papilledema, even with chronic elevations in ICP, has been addressed in numerous case descriptions. There have been many single cases and small case series documenting patients with elevated ICP who had no or unilateral papilledema. (5–12) Several have addressed the issue more systematically.

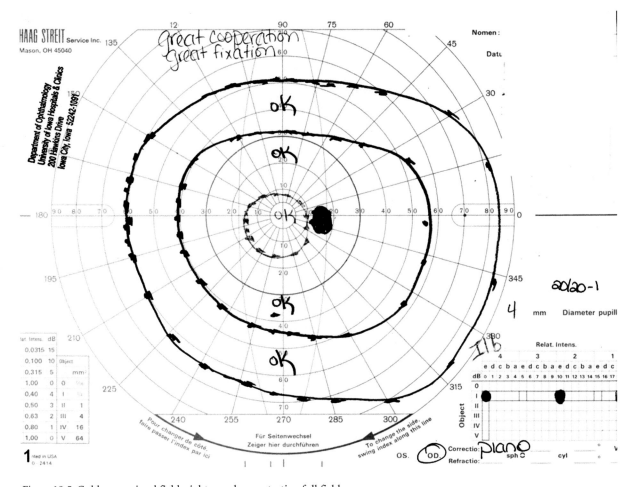

Figure 19.5 Goldmann visual field, right eye, demonstrating full field.

Vieira et al. (13) performed lumbar punctures on 60 patients with frequent headaches (classified as migraine).(13) They identified six (10%) with opening pressures > 200 mmH2O. Five of the six were overweight (body-mass-index > 29) and likely suffered from IIHWOP. Mathew et al. (14) in a spinal fluid study of 85 patients with "refractory transformed migraine", identified 12 patients with IIH who did not have papilledema.(14) In a similar study, Wang et el. identified 25 patients with chronic daily headache secondary to elevated ICP who did not have papilledema. (15) The majority of patients in this study were also overweight females and otherwise characteristic of typical IIH. More recently in 2003, Torbey et al. (16) performed continuous CSF pressure monitoring in patients suspected of having IIH without papilledema.(16) Ten patients with transient elevations in ICP were identified and assigned the diagnosis of IIHWOP. More importantly, these 10 patients had resolution of their headaches following various shunting procedures Taken together, over 50 patients have been documented in the literature to have elevated ICP who

did not develop papilledema. Most were identified during the evaluation of chronic headache suggesting that the intracranial hypertension was abnormal and did not just represent normal outliers. Moreover, the chronicity of symptoms implies that they were not simply caught during the lag period between onset of elevated ICP and the evolution of papilledema.

The question remains "Do signs or symptoms ever develop in patients with elevated ICP in the absence of papilledema?" Two well- documented cases argue that they can. Cole and George described a patient with unilateral disk edema secondary to elevated ICP. Notably the patient reported visual obscuration in both the eyes with edematous and nonedematous optic discs. (11) In a very nicely-described case, Golnik et al. (12) presented a patient with IIH, well documented to have progressive visual field loss when the optic disks were not edematous. Subsequent optic nerve sheath fenestration improved visual function in this patient.(12) This case argues strongly that elevated ICP can compromise the optic nerve in the absence of visible edema.

In closing, there is sufficient data to establish the existence of IIHWOP.(17) Patients with IIHWOP have similar demographics as those with IIH. Patients with IIHWOP may suffer from chronic headaches which respond to CSF shunting procedures. In very rare cases, transient visual obscuration and even visual loss and optic atrophy may develop in the absence of papilledema. Although a "protective membrane" located within the subarachnoid space has been postulated, this has never been confirmed and the pathophysiologic difference between these patients and those that develop papilledema remains to be determined.

REFERENCES

1. Hayreh MS, Hayreh SS. Optic disc edema in raised intracranial pressure. I. evolution and resolution. Arch Ophthalmol 1977; 95: 1237–44.
2. McCulley TJ, Lam BL, Bose S, Feuer WJ. The effect of optic disk edema on spontaneous venous pulsations. Am J Ophthalmol 2003; 135: 706–8.
3. Pagini LF. The rapid appearance of papilledema. J Neurosurg 1969; 30: 247–9.
4. Glowacky J. Fulminating papilledema. Acta Med Pol 1962; 3: 203–6.
5. Lipton HL, Michelson PE. Pseudotumor cerebri syndrome without papilledema. JAMA 1972; 220: 1591–2.
6. Sher NA, Wirtschafter J, Shapiro SK, See C, Shapiro I. Unilateral papilledema in "benign" intracranial hypertension (pseudotumor cerebri). JAMA 1983; 250: 2346–7.
7. Bono F, Messina D, Giliberto C et al. Bilateral transverse sinus stenosis and idiopathic intracranial hypertension without papilledema in chronic tension-type headache. J Neurol 2008; 255(6): 807–12.
8. Bono F, Messina D, Giliberto C et al. Bilateral transverse sinus stenosis predicts IIH without papilledema in patients with migraine. Neurology 2006; 67(3): 419–23.
9. Quattrone A, Bono F, Oliveri RL et al. Cerebral venous thrombosis and isolated intracranial hypertension without papilledema in CDH. Neurology 2001; 57(1): 31–6.
10. Marcelis J, Silberstein SD. Idiopathic intracranial hypertension without papilledema. Arch Neurol 1991; 48(4): 392–9.
11. Cole A, George ND. Unilateral papilloedema with transient visual obscurations. Eye 2006; 20(9): 1095–7.
12. Golnik KC, Devoto TM, Kersten RC, Kulwin D. Visual loss in idiopathic intracranial hypertension after resolution of papilledema. Ophthal Plast Reconstr Surg 1999; 15(6): 442–4.
13. Vieira DS, Masruha MR, Gonçalves AL et al. Idiopathic intracranial hypertension with and without papilloedema in a consecutive series of patients with chronic migraine. Cephalalgia 2008; 28(6): 609–13.
14. Mathew NT, Ravishankar K, Sanin LC. Coexistence of migraine and idiopathic intracranial hypertension without papilledema. Neurology 2996; 46: 1226–30.
15. Wang SJ, Silberstein SD, Patterson S, Young WB. Idiopathic intracranial hypertension without papilledema: a case-control study in a headache center. Neurology 1998; 51(1): 245–9.
16. Torbey MT, Geocadin RG, Razumovsky AY, Rigamonti D, Williams MA. Utility of CSF pressure monitoring to identify idiopathic intracranial hypertension without papilledema in patients with chronic daily headache. Cephalalgia 2004; 24(6): 495–502.
17. Marcelis J, Silberstein SD. Idiopathic intracranial hypertension without papilledema. Arch Neurol 1991; 48(4): 392–9.

CON: PSEUDOTUMOR CEREBRI WITHOUT PAPILLEDEMA DOES NOT EXIST

Michael Lee

From what we have been told about this patient, she does NOT yet carry a diagnosis of pseudotumor cerebri (PTC), also known as idiopathic intracranial hypertension (IIH). She must meet the modified Dandy criteria to fulfill the diagnosis of IIH. The criteria include:

1. Signs and symptoms of increased intracranial pressure (ICP)
2. Nonlocalizing neurologic examination
3. Normal MRI/MRV with no evidence of hydrocephalus, mass lesion, or venous sinus thrombosis
4. Increased opening pressure (greater > 25 cm H20) on lumbar puncture
5. Normal cerebrospinal fluid constituents
6. No other cause of increased ICP discovered

Some recognized authorities have suggested modifying these criteria further to remove the necessity of signs (papilledema) and symptoms (headache, tinnitus). As a general rule, I believe either signs OR symptoms of increased ICP should be present to make the diagnosis. It has been debated whether IIH without papilledema (IIHWOP) actually exists or not. As with all disorders, there remains a spectrum of disease and there are a few small case series and reports that adequately document what appears to be true IIHWOP.

However, IIHWOP remains a substantial minority of cases and more than likely many cases are misdiagnosed. This inaccuracy usually starts with a bias—an obese young woman with a new headache likely has IIH. The annual incidence of IIH among obese young women is 2 out of 10,000. A much higher percentage of obese women than this develop new headaches each year.

Before making a diagnosis of IIHWOP, it is important to make sure each aspect of the diagnostic criteria are fulfilled and that each is properly performed. For instance, a patient may have true papilledema, but the nerves are described as normal rendering a diagnosis of IIHWOP. There are reports in the literature of inexperienced observers using direct ophthalmoscopy to identify the lack of papilledema despite increased ICP. In general, when intracranial pressures are over 20 cm H20, spontaneous venous pulsations (SVP) disappear. Confronted with borderline optic discs, the presence of SVP suggests anomalous optic discs without papilledema. However, studies have shown that 24 hour ICP monitoring of IIH patients can vary between 50 and 500 mmHg. Other reports have described SVP in patients with elevated ICP. The presence of SVP may lead the clinician to determine the optic nerves are normal.

The misdiagnosis of IIHWOP may also result from a poorly performed opening pressure measurement during lumbar puncture. A patient with a headache may have a truly normal ICP, but the measured pressure appears falsely elevated. A properly performed opening pressure is measured with the patient in the left lateral decubitus position so that the manometer is "zeroed" at the level of the right atrium. After insertion of the needle, the patient should stretch out her neck and her legs, breathe normally, and relax. Failure to do any of these can artificially raise the measured pressure.

Increasingly, clinicians consult interventional radiologists to perform lumbar punctures. In these cases, the needle is placed under fluoroscopic guidance with the patient in the prone position. The accuracy of the opening pressure in the prone position has not been adequately studied and may not correlate accurately with the patient in the left lateral decubitus position. It is conceivable that the pannus of an obese individual against the table in the prone position could raise the intraabdominal pressure.

Finally, consider that obese patients may have headaches for an entirely different and much more common reason. Chronic daily headache occurs in up to 4% of the adult population and the risk increases with obesity. Other considerations include chronic tension headaches and chronic migraine with an overlap in peak age with IIH. Many patients that develop headaches tend to take analgesics and overuse can lead to persistently recurring (rebound) headaches.

SUMMARY

Patients with idiopathic intracranial hypertension (IIH) or pseudotumor cerebri (PTC) typically have papilledema. The modified Dandy criteria for the diagnosis of IIH however do not require papilledema. It is well established that patients with IIH can have markedly asymmetric or even frankly unilateral papilledema. This is presumably due to some structural variant in the optic nerve that does or does not allow the transmission of the increased intracranial pressure along the sheath to the involved or uninvolved disc head. In addition, patients with documented increased intracranial pressure on lumbar puncture have been reported without papilledema and patients treated for IIH have resolution of their disc edema despite repeat lumbar punctures showing persistently elevated intracranial pressures. Patients with intermittent increased intracranial pressure as seen in obstructive sleep apnea can also have normal discs. We believe that IIH can occur without papilledema but that the situation is an uncommon presentation. The clinical significance of the finding of IIH without disc edema is that visual loss should not occur in the absence of visible papilledema and if present should prompt consideration for alternative etiologies for the visual loss (e.g., inflammatory or other optic neuropathy).

20 Does a patient with an isolated vasculopathic ocular motor cranial nerve palsy need a neuroimaging study?

A 68-year-old female is complaining of new onset double vision. She does not smoke but has a past medical history of hypertension, diabetes, and hyperlipidemia. She has no past history of cancer. She noticed the horizontal diplopia yesterday only when looking at distance but today, it is worse and constant. She has no headache, no nausea or vomiting. She does not complain of any transient visual obscuration. On examination, her vision is 20/20 OU with completely normal visual fields. She has an esotropia of 40 prism diopters in primary position, with worsening of the esotropia in left gaze (Figure 20.1). She appears to be orthotropic in right gaze. Slit lamp exam was within normal limits. Dilated fundus examination revealed no optic disc edema bilaterally.

PRO: DESPITE A CLINICAL DIAGNOSIS OF ISCHEMIC CRANIAL NERVE PALSY, NEUROIMAGING SHOULD BE STRONGLY CONSIDERED

Nicholas Volpe

Isolated ocular motor palsies are acute, acquired, neurologic deficits that clearly can result from demyelination, space occupying lesions, and strokes. The fact that an ocular motor palsy is acute, painless, and isolated does not rule out the possibility of a space occupying lesion, stroke, or demyelination and for that reason every patient with an acute presentation of a cranial nerve palsy, particularly third and sixth nerve palsies, should have a neuroimaging study. Admittedly, the vast majority of patients, presenting

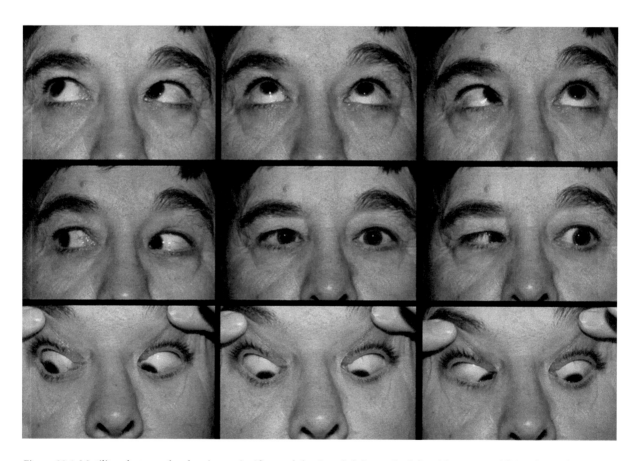

Figure 20.1 Motility photographs showing a significant abduction deficit on the left, without any additional motility disturbance.

with isolated ocular motor palsies in a vasculopathic population like the one presented are going to have negative studies and simply have their condition on the basis of a vasculopathic ischemic demyelination. This is not to say that in any population however, there would not be a small subset of patients identified with compressive lesions that are either tumors or vascular lesions, and intraparenchymal lesions such as demyelination and/or stroke. The yield of neuroimaging in this subset of patients has only been studied in a prospective fashion on one occasion by Chou et al.(1) They looked at 66 consecutive patients with isolated ocular motor palsies to determine the role of history and imaging in the diagnosis. Although they found a peripheral microvascular ischemia as an etiology in the majority of patients in this age group, other causes were identified by magnetic resonance imaging (MRI) or computed tomography (CT) scanning in 9 (14%) of patients. Diagnoses included brainstem and skull base neoplasms, brainstem infarcts, aneurysms, demyelinating disease, and pituitary apoplexy. Four of nine patients were third nerve palsy, 4 were sixth nerve palsy and only one was a fourth nerve palsy (neoplasm). One could argue that perhaps delaying such a diagnosis to the point where the patient does not recover as would be expected with a vasculopathic cranial nerve palsy and then perform neuroimaging would be reasonable. However, that would seem to be backward logic in almost any other clinical situation and cause a missed opportunity to acutely alter the patient's management. There is little doubt that isolated ocular motor palsies can be manifestations of significant intracranial disease and that the sooner that these conditions are recognized the more expeditious a patients work up can be accomplished and appropriate treatments offered. There certainly is reason to believe that a brainstem stroke that is recognized acutely would be better served in the short run with further diagnostic testing and potential treatment. It would certainly be important to not miss a demyelinating lesion that would subsequently disappear, in this era of potential immunomodulatory therapy in a middle aged patient who might have an isolated palsy as a presenting manifestation of multiple sclerosis. Finally, while the argument that there is no rush to diagnose something such as a meningioma or a cavernous sinus aneurysm in an elderly patient whose ocular motor palsy presents acutely has merit, there is certainly no disadvantage to making this diagnosis sooner rather than later to expedite work up and treatment and avoid medical liability. The traditional teaching of waiting to see if the patient improves before recommending the workup is flawed. Volpe et al. (2) showed that spontaneous resolution of sixth nerve palsy occurs in patients with significant skull base lesions.

Finally, patients are generally best served for any number of reasons by the timely diagnoses of intracranial neoplasms (effect on other cranial nerves, seizure potential), even if they are benign, so as to set forth in motion the decision making about treatment before the clinician's judgment can be called into question.

The argument to not image these patients is old and tired. It comes from an era in which only invasive, dangerous tests were available to identify neoplasms and imaging studies were insensitive for identifying strokes and demyelination. In those situations the ophthalmologist could be a "diagnostic hero," identifying isolated cranial nerve palsy and proposing, fairly safely and accurately, a vasculopathic process that would recover. In this day age there is ample availability of noninvasive testing, particularly magnetic residence imaging, that allows for an expeditious and efficient diagnosis of neoplasms, vascular lesions, demyelination, and stroke in a small segment of patients who might have otherwise been thought to have isolated vasculopathic palsies.

REFERENCES

1. Chou KL, Galetta SL, Liu GT et al. Acute ocular motor mononeuropathies: prospective study of the roles of neuroimaging and clinical assessment. J Neurol Sci 2004; 219: 35–9.
2. Volpe NJ, Lessell S. Remitting sixth nerve palsy in skull base tumors. Arch Ophthalmol 1993; 111(10): 1391–5.

CON: A CLINICAL DIAGNOSIS OF ISCHEMIC CRANIAL NERVE PALSY EXCLUDES THE NEED FOR NEUROIMAGING

Wayne T Cornblath

The sudden onset of binocular diplopia is a common complaint that presents to the ophthalmologist. Despite usually arising from a neurological cause, such as stroke, tumor or myasthenia gravis, the ophthalmologist is typically the first to see these patients and decide on the need for evaluation other than the initial examination. Of note, our discussion will apply to either transient diplopia or fixed diplopia. There are two possible choices; to image everyone with diplopia or to image selected cases. Clearly, imaging everyone with diplopia will both be very expensive and not very productive. With guidance from the literature we can decide when to image and when to observe. We will start by dividing these patients into categories. The first division will be into congenital and acquired diplopia. This would include both esotropia noted at 1 year of age and decompensation of a congenital fourth nerve palsy at age 70. The examination findings will determine whether patients fit into the congenital group, age is not a factor. This group needs no additional evaluation, only management of symptoms. The next division is by age, either age > 50 or age < 50. Most would agree that all patients < 50 years of age require evaluation including focused neuroimaging (neuroimaging directed at the area of interest, that is, cranial nerve protocol MR *versus* routine brain MR), testing for myasthenia gravis and possibly lumbar puncture. Consideration of Miller-Fisher variant of Guillain-Barre, Lambert-Eaton syndrome and other esoteric causes of diplopia can also arise. Of course, there are exceptions such as postlumbar puncture 6th nerve palsy, or cranial nerve palsy in a patient with marked vascular risk factors. The patients over age 50 with acquired diplopia can be further subdivided into cranial nerve palsy (3, 4 or 6) or brainstem abnormality based on examination. So, a skew deviation or internuclear ophthalmoplegia would fall into the brainstem category and require further evaluation, in particular focused neuroimaging. Patients over age 50 with a cranial nerve palsy are now divided into isolated or nonisolated. A diagnosis of nonisolated has two components, history and examination. A previous history of cancer, including basal and squamous cell cancers removed from the face or a history of trauma at the time of onset of diplopia would qualify as nonisolated. On examination, nonisolated patients have involvement of more than one cranial nerve, cranial nerve palsy plus other neurological symptoms (hemiparesis, ataxia, etc), proptosis, eyelid changes of thyroid eye, aberrant regeneration, or concomitant zoster. Patients with a diagnosis of nonisolated cranial nerve palsy will require additional evaluation; directed imaging, lumbar puncture, etc. Finally, patients over age 50 with an isolated cranial nerve palsy are divided into acute, subacute, or chronic. Subacute would be patients who have noted progression of double vision or development of new symptoms, that is, ptosis, over a 1 week or longer period. Chronic patients present to the physician with symptoms present for longer than 2 months. Patients over age 50 with an isolated cranial nerve palsy that is subacute or chronic require evaluation. This leaves us to contemplate what evaluation is needed for a patient over the age of 50 with an acute, isolated 3rd, 4th or 6th nerve palsy.

The first diagnostic consideration in all these patients should be giant cell arteritis (GCA). Up to 15% of patients with GCA can present with diplopia, and the diplopia can be followed by visual loss if treatment is not initiated. A careful review of systems asking for new onset headache, scalp tenderness, jaw claudication, polymyalgia rheumatica symptoms, fever, weight loss, and night sweats needs to be undertaken. In addition, Westergren erythrocyte sedimentation rate and C-reactive protein need to be drawn.(1) If the clinical index of suspicion is high corticosteroids should be started and temporal artery biopsy arranged.

The next diagnostic consideration is aneurysmal third nerve palsy, a subset of compressive lesions with a significant risk of morbidity and mortality if undiagnosed. In evaluating patients with third nerve palsy there are two components to consider, pupillary examination for anisocoria and reactivity and grading of ophthalmoplegia as complete or incomplete. Patients with anisocoria > 0.5 mm, sometimes referred to as "relative pupil sparing", need evaluation for a compressive lesion regardless of the degree of ophthalmoplegia (complete or partial).(2) Similarly, patients with complete pupil involvement need evaluation.(3) Evaluation should start with emergent MRI and MRA (although CTA is the preferred study at some institutions for the evaluation of aneurysm). This will evaluate the possibility of nonaneurysmal compressive lesions that account for up to half of compressive third nerve palsies. (4, 5) While MRA can miss some aneurysms the addition of MRA to an MRI adds only a small amount of time to the study and if an aneurysm is seen the diagnosis is made.(6, 7) If the MRI/MRA is negative the next test will depend on the institution and the clinical situation. In some institutions CTA

will eliminate the possibility of a symptomatic aneurysm but in other situations catheter angiogram will be required.(4) Clinical judgment will also be needed. In some patients with significant vascular risk factors, negative MRI and negative MRA (or CTA), mild anisocoria, and complete ophthalmoplegia clinical follow-up might be a reasonable alternative to the risk of catheter angiogram.(1) The next group is patients with a normal pupil and incomplete ophthalmoplegia. These patients can harbor an aneurysm or other compressive lesion that with additional time will develop pupil involvement.(5) Since additional time can also lead to aneurysm rupture these patients need emergent evaluation identical to those with anisocoria and partial ophthalmoplegia. Finally, patients with a normal pupil and complete ophthalmoplegia do not need evaluation for aneurysm.(2)

We are now left with three groups to consider, third nerve palsy with normal pupil and complete ophthalmoplegia, fourth nerve palsy and sixth nerve palsy. For all practical purposes these groups can be discussed as one. While several large series review cranial nerve palsies, notably the Mayo Clinic series of 4,278 cases, there are a number of difficulties with these series.(3–5) These series are retrospective, did not have a standardized evaluation protocol, cover different imaging eras (no CT, CT, MR), and include patients under the age of 50. However, there are several facts that can be quite helpful. Richards et al. (4) noted that in their review of 4,278 cases that patients with a final diagnosis of vascular cranial nerve palsy "tended to recover in 4–6 weeks".(4) In a smaller series of 221 patients 55% of sixth nerve palsies and 90% of fourth nerve palsies in patients over the age of 50 were eventually diagnosed as vascular in origin.(4) There is one prospective study that imaged all patients with isolated cranial nerve palsies.(10) We can apply our criteria to their patients for comparison purposes. There were 29 patients with third nerve palsy. Eight patients with complete ophthalmoplegia and normal pupils were evaluated and diagnosed as vascular. These are patients who would not have been imaged with our criteria above. Twenty-one patients with either pupil involvement or incomplete ophthalmoplegia were evaluated, other causes (aneurysm, brainstem stroke, tumor) were found in four. All of these patients would have been imaged with our criteria. Two neoplasms, one brainstem infarct, demyelinating disease, and pituitary apoplexy were found in the remaining patients. However, two of these five patients had symptom progression for 21 and 105 days respectively. With our criteria these two patients would have been imaged. While there is a case report titled "Sudden death from pituitary apoplexy in a patient presenting with an isolated sixth cranial nerve palsy", careful reading of the case brings out two points.(11) First, the patient's pituitary apoplexy was discovered only 6 days after his initial examination with development of new symptoms. Death occurred in the hospital 2 days after his apoplexy was known. It is not clear that an MRI ordered the day of the initial examination would have been done by 6 days, or that discovery a few days earlier would have made a difference in the final outcome. We can add a final criteria for patients who are not initially imaged, namely observation of the patient at 4 weeks. If there is no improvement, or worsening, proceed with directed imaging followed by additional evaluation if imaging is negative. The delay in diagnosis of 4 weeks in the diagnoses we are dealing with, including the remaining three patients in Chou's paper, would not be significant. In addition, the number of imaging studies saved given the percentages of patients over 50 with vascular cranial nerve palsies would be significant.

REFERENCES

1. Parikh M, Miller NR, Lee AG et al. Prevalence of a normal C-reactive protein with an elevated erythrocyte sedimentation rate in biopsy-proven giant cell arteritis. Ophthalmology 2006; 113(10): 1842–5.
2. Jacobson DM. Relative pupil-sparing third nerve palsy; etiology and clinical variables predictive of a mass. Neurology 2001; 56: 797–8.
3. Trobe JD. Isolated Pupil-sparing third nerve palsy. Ophthalmology 1985; 92: 58–61.
4. Richards BW, Jones FR Jr, Younge B. Causes and prognosis in 4,278 cases of paralysis of the oculomotor, trochlear, and abducens cranial nerves. Am J Ophthalmol 1992; 113: 489–96.
5. Lee AG, Hayman LA, Brazis PW. The evaluation of isolated third nerve palsy revisited: an update on the evolving role of magnetic resonance, computed tomography, and catheter angiography. Surv Ophthalmol 2002; 47(2): 137–57.
6. Kissel JT, Burde RM, Klingele TG, Zeiger HE. Pupil-sparing oculomotor palsies with internal carotid-posterior communicating artery aneurysms. Ann Neurol 1983; 13: 149–54.
7. Miller RW, Lee AG, Schiffman JS et al. A practice pathway for the initial diagnostic evaluation of isolated sixth cranial nerve palsies. Med Decis Making 1999; 19: 42–8.
8. Lee AG, Hayman LA, Beaver H et al. A guide to the evaluation of fourth cranial nerve palsies. Strabismus 1998; 6(4): 191–200.
9. Akagi T, Miyamoto K, Kashii S, Yoshimura N. Cause and prognosis of neurologically isolated third, fourth or sixth cranial nerve dysfunction in cases of oculomotor palsy. Jpn J Ophthalmol 2008; 52: 32–5.
10. Chou KL, Galetta SL, Liu GT et al. Acute ocular motor mononeuropathies: prospective study of the roles of neuroimaging and clinical assessment. J Neurol Sci 2004; 219: 35–9.
11. Warwar RE, Bhullar SS, Pelstring RJ, Fadell RJ. Sudden death from pituitary apoplexy in a patient presenting with an isolated sixth cranial nerve palsy. J Neuro-Ophthalmol 2006; 26: 95–7.

SUMMARY

Most patients with an isolated and presumed vasculopathic ocular motor cranial neuropathy will resolve without treatment. Some of these patients however just happen to be coincidently vasculopathic and have an underlying structural lesion producing the cranial nerve palsy. The traditional approach to these patients has been observation for improvement over time and if the palsy resolves then no neuroimaging or work up is recommended. The advent of safe and relatively easy to obtain neuroimaging however has made it possible to image all of these patients. Part of the controversy is that the medicolegal climate in certain parts of the country is such that "missing a tumor" is not an acceptable option even if the majority of isolated and presumed vasculopathic ocular motor cranial neuropathies are benign. The majority of these structural lesions (e.g., meningioma, clival chordoma) would be imaged eventually under the traditional observation paradigm because they would not resolve and might even progress during the observation period and the "delay" in diagnosis would not change the outcome in the majority of benign lesions. Nevertheless, there will be some patients who are harboring an occult structural lesion and discovering this earlier might make a difference. We recommend that whichever strategy the clinician employs (i.e., neuroimaging or not neuroimaging) that the patient be informed of the rationale for the decision making. The decision to image should also include cost effectiveness factors for an individual patient as well as the psychological implications and "make up" of the individual patient.

Index

Page references in *italics* refer to tables and figures

American Academy of Ophthalmology 83, 85
American College of Rheumatology 31
amiodarone 74
  with optic neuropathy
    causes of 74–6
    criterias 77–8
    time order constraint 75
    variation 74–5
  without optic neuropathy 76–8
aneurysms 65–6
  diagnosis of 64
angiotensin-converting enzyme 6
anisocoria 102
anterior ischemic optic neuropathy
  erectile dyfunction agents, causes 67
antiphospholipid antibody syndrome 47
antiplatelet therapy 24, 25
apraclonidine 88
arteritic AION 91
aspirin 47
azathioprine 83

Bartonella henselae 6, 15
  treatment for 15
bilateral temporal artery biopsy 29–31
biopsy 29

cardiac echo 45
carotid Doppler 45
cat scratch disease 14
  treatment 15
cat scratch titer 6
cerebral transient ischemic attacks
  (TIAs) 46
cerebral venous sinus thrombosis (CVST) 43
cerebrospinal fluid (CSF) 9
cerebrovascular disease 59
chronic obstructive sleep apnea 72
Cialis 72
clinically definite multiple sclerosis (CDMS) 12
  interferon beta therapy 12
closed head injury (CHI) 38
  high-dose steroid 38
cocaine 88
Combined Immunosuppression and Radiotherapy in Thyroid
  Eye Disease (CIRTED) 83

computed tomography angiography (CTA) 64
  see also negative computed tomography angiography (CTA)
Corticosteroid Randomization After Significant Head injury
  (CRASH) trial 36, 38
corticosteroids 12
CT venography (CTV)
  pseudotumor cerebri 41, 44

Dandy criteria 43, 98
  idiopathic intracranial hypertension
    diagnosis of 99
Devic's Syndrome 9
diplopia 101
disc pallor 6
dominant optic atrophy (OPA1) 6

erectile dyfunction agents
  with anterior ischemic optic neuropathy 67–71
  without anterior ischemic optic neuropathy 71
  NAION
    causes of 71–2
    treatment of 68
euthyroid Graves disease 84

fluorescein angiography
  with AION 92–3
  with giant cell arteritis (GCA) 91–3
    case example 92
  without giant cell arteritis (GCA) 93
fluorescent Treponemal antibody (FTA-ABS) 6
Food and Drug Administration 69

generalized myasthenia gravis (GMG) 81
giant cell arteritis (GCA) 30, 91–3, 102
  antiplatelet therapy 26–7
  biopsy 31–2
  case example 92
  criteria for 31
  diagnosis of 31
  IV steroids 24, 25–6
  malpractice 33
  oral steroids 28
  potential sequelae 28
  symptoms 93
  systemic manifestations of 93
  treatment for 24

Goldmann (kinetic) perimetry 53
  and Humphrey perimetry 52–3
  nonspecific findings 54, *55*
  for visual field test 51–4

high-dose steroid 35
  in closed head injury (CHI) 38
  homonymous hemianopsia
  cerebrovascular disease 59
  visual rehabilitation therapy 59–61
    advantages 59–61
    disadvantages 62
Horner's syndrome 87
  in adult population 89
  cause of 87
    structural lesion 88
  diagnosis of 87
  neuroimaging 89
  pharmacological testing 87
Humphrey perimetry 52–3
hydroxyamphetamine 88
hypercholesterolemia 21
hyper-coagulation 72
hyperemia 94
hyperthyroidism 82

idiopathic intracranial hypertension (IIH)
    40, 43, 96, 98
  diagnosis of 98–9
  misdiagnosis of 99
  without papilledema (IIHWOP)
    diagnosis 98–9
    misdiagnosis 99
idiopathic neuroretinitis 18
intracranial pressure (ICP) 94
  papilledema
    absence of 97
iris tears 4
ischemic cranial nerve palsy 100
  clinical diagnosis
    advantage 100–1
    disadvantage 101–3
ischemic optic neuropathy decompression
    trial (IONDT) 20
isolated ocular motor palsy 100
isolated optic atrophy
  acute visual loss symptoms 1
  laboratory evaluation
    advantage 1–4
    disadvantage 4–6
  neuroimaging
    advantage 1–4
    disadvantage 4–6

neuropathologies
    treatable causes for 4
  patients with
  IV steroids 24, 25–6

Koch's postulates
  for causal association 74

laboratory evaluation 1–6
Leber's hereditary optic neuropathy 6
Levitra 72
low dose oral aspirin therapy 28
lupus erythematosus 47
Lyme titer 6
lyme 13

macular star 18
magnetic resonance angiography (MRA) 64
  see also negative magnetic resonance angiography (MRA) 63
magnetic resonance image (MRI) scan 1, 40
magnetic resonance venogram (MRV) 40
methylprednisolone 28
miosis 89
multiple sclerosis 7
  testing 7, 11
  treatment 7, 11
myasthenia gravis
  diagnosis 79
  prevention 79
    side effects 81
  steroids 79
  treatment 79

National Acute Spinal Cord Injury Studies (NASCIS) 38
negative computed tomography angiography (CTA) 63
  adequacy of 65–6
  with third nerve palsy 63–4
negative magnetic resonance angiography (MRA) 63
  adequacy of 65–6
  with third nerve palsy 63–4
neuroimaging 1–6, 100
  Horner's syndrome 89
  ischemic cranial nerve palsy
    pros 100–1
    cons 101
Neuromyelits Optic (NMO) 9
neuroretinitis patients
  causes 14
  differential diagnosis 18
  ocular manifestations 15
nonarteritic anterior ischemic optic neuropathy (NAION) 6,
    67, 69, 71–2, 76–7
  blood pressure

24 hour blood pressure monitoring 20, 22
  nocturnal hypotension 19–20, 22
erectile dyfunction agents 67–71
hypercholesterolemia 21, 22
hypotension 73
risk factors 72
sleep apnea syndrome (SAS) 20, 2
tobacco 21, 22
vasculopathic patient 19
NovaVision AG 62

ocular myasthenia gravis (OMG) 81
optic disc edema 18
optic nerve atrophy 1
optic neuritis (ON)
  evaluation and management 7
  recovery 8
  with multiple sclerosis 8
  testing 11–12
  treatment 12
Optic Neuritis Treatment Trial (ONTT) 8
optic neuropathy
  with amiodarone 74
optokinetic nystagmus (OKN) therapy 59
oral placebo 83
oral prednisone 24, 25
oral steroids 79–80
  for giant cell arteritis (GCA) 28

pallor 1
Pancoast tumor 88
papilledema
papilledema 94, 96
  absence of
    idiopathic intracranial hypertension 94–8
    pseudotumor cerebri 98–9
  in sinus thrombosis 41
phosphodiesterase-5 (PDE-5) inhibitors 72
posterior communicating (PComm) artery 65
posterior ischemic optic neuropathy (PION) 93
prednisone 28
pregnancy 44
pseudotumor cerebri 40, 43, 94
  diagnosis of 43
  IIHWOP 94–8
  magnetic resonance imaging (MRI) 40, 41–2
    disadvantages 43
  magnetic resonance venogram (MRV) 40, 41–2
    disadvantages 43
  without papilledema 98–9
    Dandy criteria 98
ptosis 89

radiation therapy 82
  efficacy of 83
  with thyroid ophthalmopathy 82–4
  without thyroid ophthalmopathy 84–5
radioactive iodine (RAI)
  with thyroid ophthalmopathy 84
Reid Longmuir 82–
Reinhardt L 11
retinal migraine 49
retinal nerve fiber layer (RNFL) 94
retinal vasospasm 47
Ruth H Schmidtke 11

shrinkage 32
sildenafil 72
sinus thrombosis
  papilledema in 41
sleep apnea syndrome (SAS) 20
Snellen visual acuity 25, 26
spontaneous venous pulsations (SVPs) 94
steroids 28
  giant cell arteritis (GCA) 24
  in myasthenia gravis
    prevention 79
    side effects 81
stroke 49, 50
suspected giant cell arteritis
  bilateral temporal artery biopsy 29–31
  unilateral temporal artery biopsy 31–3
  see also giant cell arteritis (GCA)
syphyllis 13

tadalafil 72
temporal artery biopsy 29–33
  clinical criteria 31
third nerve palsy 102
  with negative MRA 63–64
  with negative CTA 63, 64
thrombophilia 44
thyroid ophthalmopathy 82
  complications 82
  evaluations of 83
  radiation therapy in
    efficacy of 82–3
  signs and symptoms 84–5
  smoking cigarettes 84
trans-esophageal echocardiogram (TEE) 47, 49
transient ischemic attack (TIA) 50
transient monocular visual loss 47
transient vision loss 47
  cardiovascular risk factor assessment 49–50
  by thromboembolic 45–9

transthoracic echography (TTE) 49
traumatic optic neuropathy
  presumed mechanism 36
  treatment of 36
    high-dose steroid 35
    surgery 38
  untreated prognosis of 38
tuberculosis (TB) 13

ultimate skip lesion 32
  see also shrinkage 32
unilateral temporal artery biopsy 31–3

vardenafil 72
vasculopathic patient
  nonarteritic anterior ischemic optic neuropathy 19–21

Viagra 72
Vision Restoration Therapy 62
visual field defect 1
visual field test
  Goldmann (kinetic) (GVF) 51–4
  homonymous *56, 57*
  Humphrey (automated) (HVF) 54–8
visual rehabilitation therapy (VRT) 60
  homonymous hemianopsia 59–61
    advantages 59–61
    disadvantages 62

Westergren erythrocyte sedimentation rate 102